THE YOUNG AND THE DIGITAL

The Young and the Digital

What the Migration to Social-Network Sites, Games, and Anytime, Anywhere Media Means for Our Future

S. Craig Watkins

BEACON PRESS
BOSTON

Beacon Press
25 Beacon Street
Boston, Massachusetts 02108-2892
www.beacon.org

Beacon Press books
are published under the auspices of
the Unitarian Universalist Association of Congregations.

13   12   11   10   8   7   6   5   4   3   2   1

This book is printed on acid-free paper that meets the uncoated paper
ANSI/NISO specifications for permanence as revised in 1992.

Text design and composition by Wilsted & Taylor Publishing Services

Library of Congress Cataloging-in-Publication Data
Watkins, S. Craig (Samuel Craig)
  The young and the digital : what the migration to social-network sites, games, and
anytime, anywhere media means for our future / S. Craig Watkins.
      p. cm.
  Includes bibliographical references and index.
  ISBN 978-0-8070-0616-0 (paperback : alk. paper)
   1. Mass media and youth—United States. 2. Technology and youth—United States.
3. Digital media—Social aspects—United States. 4. Teenagers—Social networks—
United States. 5. Internet—Social aspects. 6. Online social networks—United States.
I. Title.
  HQ799.2.M352W37 2009
  303.48'3308350973—dc22   2009010400

Many names of people mentioned in this work have been changed to protect their
identities.

To Cameron Grace Watkins,
daughter, friend, digital native, and fellow author

CONTENTS

INTRODUCTION The Young and the Digital ix

ONE Digital Migration:
Young People's Historic Move to the Online World 1

TWO Social Media 101:
What Schools Are Learning about
Themselves and Young Technology Users 19

THREE The Very Well Connected:
Friending, Bonding, and Community in the Digital Age 47

FOUR Digital Gates: How Race and
Class Distinctions Are Shaping the Digital World 75

FIVE We Play: The Allure of Social Games,
Synthetic Worlds, and Second Lives 103

SIX Hooked: Rethinking the
Internet Addiction Debate 133

SEVEN Now! Fast Entertainment and
Multitasking in an Always-On World 157

EIGHT "May I have your attention?":
The Consequences of Anytime, Anywhere Technology 171

CONCLUSION A Message from Barack:
What the Young and the Digital
Means for Our Political Future 193

The Making of This Book:
Research, Methods, and Acknowledgments 209

Notes 219

Index 243

The Young and the Digital

[Facebook is] a big part of our lives . . . in this day and age.
And if you're not a part of that, then you're missing a huge
part of your friends' lives also.

—Erica, twenty-two-year-old college student

In 2004, Rupert M. Murdoch, chairman of the third-largest media con-
glomeration in the world, News Corporation, saw the future—and it
was digital. Murdoch's life in the media business began in 1953 with
the purchase of a small newspaper in his native Australia. Over the span
of his legendary career, Murdoch built a fortune and a global media
empire by buying up print and television properties all across Europe,
Asia, and North America. But Murdoch's empire was showing signs of
aging. New media technologies like the Internet were eroding the once
taken-for-granted power of the old media guard. The print business, for
instance, was steadily decreasing in value as digital content and citizen
journalism ascended. Television too was appearing more like a relic in
the age of interactive and user-generated media. In a Washington, D.C.,
speech before the American Society of Newspaper Editors, Murdoch
openly acknowledged his troubles.

"I didn't do as much as I should have after the excitement of the late
1990s," he told the gathering. "I suspect many of you in this room did
the same, quietly hoping this thing called the digital revolution would
just limp along. Well it hasn't, it won't, and it's a fast-developing reality
that we should grab."

The man *Time* magazine called "The Last Media Mogul" and the
self-described "digital immigrant" spoke like a technology convert. "We
may never become true digital natives but we can and must begin to
assimilate to their culture and way of thinking." In a tone that matched

the urgency of the moment, Murdoch added, "It is a monumental, once-in-a-generation opportunity . . . if we are successful our industry has the potential to reshape itself and to be healthier than ever before."

According to insiders at News Corporation, it was a 2004 Carnegie Corporation survey of eighteen- to thirty-four-year-olds that galvanized Murdoch. The study found that 44 percent of the respondents use an Internet portal at least once a day, whereas 19 percent reported reading a newspaper. Among the youngest respondents in the study, 10 percent believed newspapers were trustworthy, and a miniscule 4 percent thought they were entertaining. In a summary of the report titled *Abandoning the News*, Merrill Brown writes, "This audience, the future news consumers and leaders of a complex, modern society, are abandoning the news as we've known it, and it's increasingly clear that a great number of them will never return to daily newspapers and the national broadcast news programs."

Carnegie's core finding—that newspaper readers were a dying breed —was a wake-up call for Murdoch. The report clearly suggested that part of Murdoch's life work and legacy, the print empire he built at News Corporation, was about to be made obsolete in the midst of great social and technological upheaval. In the next ten years the company, in its current state, would be culturally irrelevant and economically vulnerable in a world gone digital. More than anything, the Carnegie Corporation survey convinced Murdoch that News Corporation needed to get young and digital . . . and fast.

In 2005 Murdoch made his move. He summoned top News Corporation executives from around the world to New York in February. Their mission was simple: craft a plan that would take News Corporation into the digital future. On July 15 the company announced the creation of a new unit they called Fox Interactive Media (FIM) to, among other things, coordinate News Corporation's Internet properties into a cohesive plan and robust digital media experience. Members of FIM acknowledged that while the media giant was well suited to attract older media users, its current structure was ill suited for young media users who were making what MIT Professor Nicholas Negroponte called "being digital" an ordinary way of life.

On August 9 News Corporation acquired IGN Entertainment, Inc., a gaming, film, and male lifestyle portal, for $680 million. It was the company's third Internet acquisition in three months for a combined total of $1.3 billion. But the real prize was a deal News Corporation revealed on July 18, a mere three days after officially announcing the creation of FIM.

Once News Corporation settled on a short list of Internet properties to consider acquiring, one site quickly rose to the top of their list. For several weeks the global media conglomerate had been in serious talks to buy the Los Angeles–based Intermix Media. Intermix owned some thirty dot.coms, but the crown jewel was the Santa Monica–based social-network site MySpace. As FIM entered into negotiations with Intermix, what they saw in MySpace was the future of the Internet and the future of media—bottom up and user driven rather than top down and corporate driven.

In 2005 the fastest-growing Web sites, according to Nielsen// NetRatings, were the ones young people, ages twelve to twenty-four, were most likely to visit. Many of these sites had several things in common: they allowed users to create and share content, while also providing strong social and community-based features. Between November 2004 and the same period in 2005 the number of unique visitors to MySpace expanded from 2,874,000 to 24,495,000. That was a massive gain of 752 percent. During the same period, Facebook grew from 932,000 to 5,869,000 unique visitors for an increase of 530 percent.

Equally important was the age of MySpace's core users. For the period of the negotiations, 25 percent of MySpace users were between the ages of twelve and seventeen and 20 percent fell between the ages of eighteen and twenty-four. This was the segment that News Corporation vigorously pursued, the segment that, according to the Pew Internet & American Life Project, was leading the transition to digital. As Ross Levinsohn, President of FIM at the time, explained, "From a demographic standpoint, we lacked the ability to reach the 15-to-30 year old. [News Corporation] as a company does well at reaching the 30-plus, but we were lacking the youth brand."

After MySpace was released in Beta form in September 2003, the site officially opened in February 2004. That same month MySpace surpassed Friendster as the number one ranked social-network site and never looked back. As the story goes, the rise of MySpace was totally viral. There were no ads, hip marketing campaigns, or corporate hype. The site benefited from the best and cheapest source of advertising, word of mouth. People joined MySpace because their friends joined. And yet the rise of MySpace was not entirely fortuitous. There was, at least initially, some degree of manufactured buzz involved.

After launching MySpace, founders Tom Anderson and Chris DeWolfe hit the trendy Los Angeles night scene. A few years earlier, Anderson was part of a San Francisco–based indie rock band named Swank. Though the band never enjoyed huge success, Anderson's familiarity with the music scene proved extremely helpful in the creation of MySpace. The site's creators made music a central part of the user experience. Bands showcased their music while fans created communities around their interests in music. Along with bands, Anderson and DeWolfe invited club goers and models to use the site. DeWolfe compared starting MySpace to opening a bar. From the beginning the site combined key elements and sensibilities—stylish-looking people and edgy bands—that made it a desirable destination for young online social-networking enthusiasts and wannabe celebrities.

"Since we were telling people in clubs—models—suddenly everyone on MySpace looks really pretty," recalled Anderson. "That wasn't really the plan. It just kind of happened."

But MySpace swiftly became more than aspiring rock stars and attractive people. Within a few short months it emerged as the online destination of choice for wired teens. How did the site come from nowhere in 2003 to be one of the world's most visited online destinations by 2005? Some called it luck. Others complained that Anderson and DeWolfe cashed in on an idea that had been percolating in Silicon Valley for years: the belief that the Web was an ideal place for connecting people to people. As DeWolfe explained, this was the core idea behind MySpace: "People are starting to understand that the holy grail of the Internet is community. The real potential for the Internet that we were talking about 10 years ago is just now beginning to materialize."

MySpace was not the first attempt to execute this vision, but it was one of the most successful.

The simple genius of MySpace was the cofounders' decision to build a one-stop destination that combined a suite of applications that appealed to young technology users. For starters, MySpace users were able to customize their own personal home pages that were an established part of the Web. The platform's Instant Message capabilities, in the tradition of America Online Instant Messenger, permitted rapid communication between users. The blogging features anticipated young people's desire to talk about anything and everything in the digital public sphere. MySpace's digital imaging features exploited the appeal of photo hosting applications that make it easy to file and share pictures. And by tapping into music, the creators of the site incorporated an enduring theme of youth culture. Once the site was built the staff did one last thing: they turned it over to users.

News Corporation's acquisition of MySpace was the social Web's big bang. In the immediate aftermath of the great dot.com bust in 2000, the major media companies grew cautious about the Web. Murdoch openly bragged that he spent a fraction of the money his rivals spent. The most celebrated failure, of course, was the ill-fated merger between America Online and Time Warner. The MySpace deal, however, was a real game changer. "The Internet is exciting again and once again folks are rushing in," proclaimed Web 2.0 evangelist and entrepreneur John Battelle in a *New York Times* editorial. The dollar amount—$580 million—was huge, but the message the deal signaled was even bigger: corporate media could no longer ignore what young people were doing online.

Much of the buzz in the press after the MySpace deal emphasized the millions of young people on the site, the rise in online advertising, and the race to monetize the social Web. In the end, though, News Corporation's determination to become a player in the digital media space was about more than delivering young eyeballs to an online advertising market that is expected to grow from $9.3 billion in 2004 to $18.9 billion in 2010, according to Jupiter Research. The global media giant was not only vying for a greater presence on the Web; it was also vying for a greater presence in the lives of the young and the digital. Like all hot pop culture brands, MySpace predictably cooled off. What the social-network site represented, though, was much more enduring than the platform. MySpace's strength, ultimately, was not in what was presumably original about it but, rather, what was quite durable about it. MySpace did not tip because it was radically new. MySpace tipped because it was radically true to what young people have always loved to do: spend time with their peers, while also expressing themselves through a music, style, language, and pop culture experience that is all their own.

About a year after the acquisition of MySpace, I went into the field with a small research team and began talking with young people about the role of technology in their lives. Our approach was also influenced by my involvement in the MacArthur Foundation's digital media and learning initiative. Our goal was simple: to understand young people's emphatic embrace of social and mobile media technologies. We collected surveys, conducted one-on-one interviews and focus groups, and visited the places—home, school, and online spaces—where young people spend the bulk of their time. It was a full-scale immersion into what I call the digital trenches. As we emerged from the trenches, our perspective on young people's engagement with new media technologies grew more detailed and dynamic.

In the pages that follow, you'll find an intimate and evidence-based portrait of a generation we simply call the young and the digital. Before our work began, I was fully aware that teens and young twenty-somethings were shifting much of their leisure activities online. But as we began to analyze the data from the surveys we collected and transcribe the stories young people shared with us about their technology-rich lives, it was vividly clear that digital is more than the tools and technology they use—it is, quite frankly, a way of life. This was especially clear in the rise of social-network sites like MySpace and Facebook.

In a relatively short time period, going online to communicate with friends, classmates, and acquaintances has become *the* activity of choice for most teens and young twenty-somethings. As Erica, a twenty-one-year-old college student, told us, "I only know two people who are not on Facebook." And she is right. Her generation is immersed in a world of social-networking enthusiasts. Ninety-seven percent of the people we survey have a personal profile on a social-network site. More than half, 58 percent, subscribe to at least two or more sites. In fact, many young people describe social-network sites as a routine part of their daily lives and can barely recall what life was like without them. Young people use social-network sites to manage many parts of their lives, including, for example, their day, their personal and professional ties, their schoolwork, and also to communicate, forge cultural identities, and connect to their world and the people around them.

Most people under the age of twenty-five cannot imagine life without connected communities. As Melinda, a twenty-one-year-old aspiring film producer, proclaimed during our conversation with her, "What do people do without Facebook?"

Three-quarters of the people in our survey visit a social-networking site at least once a day. Half of them visit three or more times a day. And a growing number of them can identify with the unabashed claims of twenty-four-year-old Kendra.

"Oh my God," she shrieked during our one-on-one with her, "I am on MySpace all day!" Kendra elaborates on the intensity of her engagement. "I'm not ashamed to admit it. It is the first thing I check when I get on the computer, the last thing I check when I get off, and the most frequented site in between."

As you might expect, people are drawn to computer-mediated social networks for a variety of reasons. Occasionally, we come across those who feel a certain degree of peer pressure to use social-network sites. The primary motivation in instances like these is the fear of becoming socially alienated.

"It seems like everyone uses it [Facebook] to do things now," says Renee, a twenty-three-year-old college graduate. "What would I be missing if I weren't on it?"

Twenty-two-year-old Dale also admitted that he "felt a lot of peer pressure to get on [MySpace] and make a profile." Dale says, "Right before I joined, it had been sweeping the nation, just before it swept the world." Initially, he resisted setting up a profile. Several of his friends were on MySpace, but he noticed how it consumed much of their time and energy. Speaking somewhat jokingly, and as if MySpace were a drug, he explained, "I knew that if I started using I wouldn't be able to stop using it. I would be just like everyone else and be on it like fifteen hours a day." Dale, in his words, did not want to "turn into a MySpace nerd." Eventually he gave in. "I was rusty at first but then once I got the nuances down, it was really plug-in and play."

Kevin, a talkative nineteen-year-old, shared a similar story with us. Like Dale, he deliberately stayed away from the whole social-networking craze. "I told myself I would never do it. I was totally against MySpace and Xanga." He thought that all of the pages looked too much alike and that his time online could be better spent with things he was truly passionate about, namely games and Web comics. But when he arrived at college, all of his dorm mates and college buddies were on Facebook and it became difficult to keep up with the online conversations and community they shared. Then, after just a couple of weeks of class, he had a generational epiphany.

"I quickly realized if you don't have a Facebook [page], it's like not having a cell phone. And if you don't have a phone, you are totally cut off from the people around you." After a few seconds of self-reflection, Kevin added, "It sounds really shallow, come to think of it. What has our society turned into?"

Still, most young people we meet were eager to set up an online profile and connect to their friends. They entered computer-mediated communities because they were curious and excited about the opportunity to extend their off-line relationships into the online environment. "I did not feel obligated to do it [join a social-networking site]; I did it because I wanted to," said Sidney, a twenty-six-year-old paralegal we met. Like many of the young people we spoke with, Sidney was already using the Web for social communication before the arrival of sites like Friendster, MySpace, and Facebook. Most of the young people we meet have a history with e-mail, chat rooms, and Instant Messenger and consider their use of social-network sites a logical progression.

But it was after a conversation with twenty-two-year-old Erica that I began to comprehend the true pull and power of social-network sites. Her motivation for getting involved in online "friending" was simple, profound, and representative of the cultural sensibilities that made social media a generational touchstone.

"It's a big part of our lives in this day and age," she said candidly during a conversation outside a small café. "And if you're not a part of that, then you're missing a huge part of your friends' lives also." Erica elaborated. "It's hard to relate to the people that you are friends with if they have this big force in their lives and you are not a part of it . . . it's the impact that it has on real life."

Like most of the young people we met, Erica moved online to do what I call life-sharing. Social and mobile media platforms have emerged as the dominant technologies in young people's lives because they offer something that television never has and never will: the constant opportunity to connect and share their lives with close friends and acquaintances. As much as anything else, social and mobile media platforms enable young people to share stories about their day, their mood, and their lives through a wide array of digital content such as pictures, blogs, small messages, and video. Sharing our lives with others via the Internet and mobile phones means we are constantly connected, accessible, social, and, sometimes, vulnerable. Life sharing, in the end, is as much about community as it is the individual.

In the pages that follow you will meet a number of people, including teens, college students, parents, and educators. Among them is a group that I call the four-pack, that is, four young men who let me follow them closely for about six months. I visited them in their dorm and spoke with them on many different occasions. They also kept very intimate diaries that charted the media they used, as well as their views about technology. By any definition, the four-pack—Brad, Derrick, Chase, and Trevor—are gamers. Their leisure activity of choice is games, and during a typical week each spends about fifteen to twenty-five hours playing them. Indeed, games are the common denominator in their strongest and most meaningful social ties.

In the past, society has tended to think of games as antisocial. Violent games, critics claim, make young people aggressive and less caring of others. The classic stereotype of gamers—glued to a screen, skillfully manipulating a controller, and socially isolated from peers—still prevails. But as I got to know the four-pack, I realized that games play an extraordinarily social role in their lives. Much like users of social-network sites, these young men were drawn to games, in part, as a way to bond with their friends.

Though Microsoft's Xbox 360 was the console of choice among the four-pack, they also enjoyed Nintendo's Wii, a more modest machine. Referring to the Wii, Brad says, "So many of the games are centered around playing with multiple people." The social aspects of the system made it a favorite in the eyes of Brad and his buddies. In the dorm, where the four-pack live, residents even got together and hosted a Wii bowling tournament, complete with a big projection screen and brackets. About forty people entered what evolved into a full-fledged sporting event with a cheering section and even an announcer. It was a community event and a big party. Laughing about the whole experience, Brad says, "It was getting pretty ridiculous toward the end, with people cheering and pulling for their favorite bowler." But this is exactly what social games and social media do—they bring people together.

Brad's description of the virtual bowling tournament is striking when you consider the current debates about the alleged threats that new communication technologies pose to the health and vitality of informal public life and community in millennial America. As I listened to Brad talk about the bowling tournament, I considered the well-traveled argument made by Harvard political scientist Robert Putnam in his book *Bowling Alone: The Collapse and Revival of American Community*. Citing declines in, among other things, bowling leagues, civic participation, and volunteerism, Putnam argues that Americans, starting around 1970, began retreating from informal public life and into the privacy of their own homes. One of the main culprits of this declining sense of community among Americans, Putnam argues, is the growth of media in our homes.

Throughout the latter half of the twentieth century, no media was more dominant in our lives than television. The debates about the impact of television content—especially sex and violence—are familiar to most. The debates about the impact of television viewing as an activity—where we watch, how often, with whom, and the social consequences—are not as familiar. Among researchers who study television viewing as a leisure activity, the medium's greatest legacy is how it influences our connection, or lack thereof, to our neighbors, communities, and the world around us. According to Putnam, "TV watching comes at the expense of nearly every social activity outside the home, especially social gatherings and informal conversations." The end result, Putnam asserts, is the erosion of social capital or the sense of neighborliness, mutual trust, and reciprocity that binds people together and strengthens community.

Over the years the spread of communication technologies throughout our homes has turned them into private hubs of information, communication, and nonstop entertainment. Ray Oldenburg, a sociologist, goes so far as to claim that the most predictable social consequence of technological innovation is that humans will grow further apart from each other. Discussing the decline of informal public life in America, Oldenburg writes, "Increasingly, her citizens are encouraged to find their relaxation, entertainment, companionship, even safety, almost entirely within the privacy of homes that have become more a retreat from society than a connection to it." This outlook, developed long before home and ubiquitous computing, persists today. But is the shift to screens that are more personal, social, and mobile really making us less connected to each other?

Conventional wisdom has long maintained that time spent at home with television is time spent away from friends and public life. And yet, computer and mobile-phone screens represent very different kinds of experiences than the ones traditionally offered by television. A great irony of life on the computer screen is the fact that we usually go online alone but often with the intent of communicating with other people. Among the teens and young adults that we talk to, time spent in front of a computer screen is rarely, if ever, considered time spent alone. Rather, the engagement with technology is viewed as an opportunity to connect with friends. Granted, connecting via a mobile phone or Facebook is a different way of bonding, but, as I argue in the following pages, these practices are expressions of intimacy and community.

Another common view is that the new media ecology is turning young people into a herd of social recluses more interested in the gadgets they own than the people in their lives. And it is true—young people *are* enthusiastic about the technology they use. But what may surprise you is that they are equally enthusiastic about their friends and acquaintances. The Wii bowling tournament is a great example of how social media platforms can bring people together, rather than pull them apart. No matter where they are—sitting in class, driving a car, or holding a face-to-face conversation—the twenty-five-and-younger set are constantly using a screen to connect to their peers.

Digital media also has its limits. All of the euphoria about social and mobile media notwithstanding, it does not seem to be bringing people together across the long-standing barriers of race and class. In 1996, two MIT scholars, Marshall Van Alstyne and Erik Brynjolfsson, published a paper investigating whether or not the electronic communities forming on the Web represent a global village or socially exclusive forms of behavior. In their paper, Van Alstyne and Brynjolfsson maintain that increased connectivity has the *potential* to create diverse communities by providing individuals the opportunity to come together across social as well as geographical boundaries. A less sanguine outcome strikes many others as a very real possibility. Acknowledging this outcome, Van Alstyne and Brynjolfsson write, "Just as separation in physical space, or basic balkanization, can divide geographic groups, we find that separation in virtual space, or 'cyberbalkanization,' can divide special interest groups."

The more we learn about teens and young twenty-somethings' use of social and mobile media, the more complex and fascinating the story becomes. College students' daily use of Facebook, for instance, reveals an intriguing paradox. Facebook makes it possible for the young people we meet to connect with each other "all day, every day." And that is what they are doing—practicing unprecedented degrees of community as they maintain constant connectivity. Still, even as their use of Facebook brings them together in extraordinary ways, it also reinforces their social and geographical distance from "others," especially MySpace users who they commonly view as unsophisticated, uneducated, and undesirable. Oddly enough, the way they use new media technologies creates both fascinating expressions of community and disheartening instances of what we call digital gating—the maintenance of social and geographical boundaries.

The move to more personal, social, and mobile screens marks a pivotal moment in our cultural history. Importantly, it signals the fall of one screen, television, and the rise of another screen, the computer, as well as notably different ways of being in the world.

In 2000 the number of households with a computer crossed the 50 percent threshold in America. Roughly four years later more than half of American households had access to the Internet. There were no celebrations or fancy pronouncements; nevertheless, the steady adoption of the networked computer ushered American life into a whole new frontier. An assortment of unique properties—its speed, availability, and the intensity and breadth of the information accessible online—makes the networked computer the most compelling household screen ever. It is also the most versatile screen to settle in our homes. We use it for leisure, work, learning, playing, shopping, bonding, civic engagement, and personal expression.

Researchers have gone to great lengths over the years to understand how social and ubiquitous computing has changed our lives—in some cases for the better and in other instances for the worse. The story, much like life itself, is rich, complex, and defies simple answers. In the following pages, I tell parts of the story, primarily from the perspectives of the young and the digital, the first generation of Americans who will spend most of their lives in a world where personal, social, and mobile media are common, widespread, and fully integrated into our daily lives.

ONE

Digital Migration

Young People's Historic
Move to the Online World

The enduring American love affairs with the automobile
and the television set are now being transformed into a giddy
passion for the personal computer. . . . Above all, it is the
end result of a technological revolution that . . . is now, quite
literally, hitting home. Americans . . . expect that in the fairly
near future, home computers will be as commonplace as
television sets or dishwashers.

> —from the 1982 "Man of the Year" issue of *Time*,
> which selected a machine: the computer

In the three decades following the end of World War II, households in
America were relatively simple places in terms of communication technology. Most American households had a radio, a record player, and a
phone line. Throughout the second half of the twentieth century, one
element was definite about communication technology in the majority
of American homes: television was king.

In 1950, 10 percent of American households had a television set.
Ten years later virtually every American home had at least one television, making the technology one of the fastest and most widely adopted
technological innovations in modern American life. Thirty years after it
was first adopted, television was still the marquee technology in American homes. It was our primary source of entertainment and information. In essence, it was our window to the world.

Throughout the 1980s Americans began to upgrade the technological capabilities of their homes at an astonishing rate. Between 1980

and 1990 multitelevision set households in America increased from 50 percent to 65 percent. During that same period cable television grew from 20 to 56 percent and the average number of channels per household went from nineteen in 1985 to about forty in 1993. In 1980 few American homes contained a video-cassette recorder, but by 1990, 66 percent of American homes had added one. Introduced in 1983, CDs went on to all but extinguish the LP album format by the close of the decade. Video-game consoles grew steadily throughout the decade and were soon a staple in many households with children and young males. During this time, communication technologies in the home not only became more abundant, they also became more individualized. For the first time in American history many school-age children gained access to their own television, phone line, gaming console, and music media, turning their bedrooms and playrooms into an oasis of media, entertainment, and private leisure.

Since the 1990s American homes have been in a perpetual state of technology upgrade, adding, among other things, satellite dishes, fax machines, high-powered gaming equipment, personal computers, the Internet, DVD players, digital cameras, MP3 players, and digital video recorders. Today, a steadily growing number of Americans live in technology-rich households, what communication scholar Jorge Schement calls the "wired castle." At the heart of the wired castle is the networked personal computer or, more precisely, the Internet.

In a rising number of homes the computer is becoming the focal point for both leisure and work-related activities, making it more versatile and, in many respects, more integral to daily household life than television. Initially, computers tended to be in work-related spaces such as home offices or in general-use areas like family rooms. The diffusion of wireless, however, turns every room—the kitchen, bedroom, living room—into a home-computing space.

The rise and influence of the personal computer is nothing short of phenomenal. Not that long ago, home computers did not exist. In 1985, when the Current Population Survey, a division of the U.S. Census Bureau, began measuring computer adoption rates, only 8 percent of American homes possessed them and the Internet was a technological experiment confined to the military and universities. By comparison, 98 percent of American homes had a television in 1985; more than half, 57 percent, had multiple sets. Gradually, as the cost of computers fell and interest in them rose, more people began adopting the technology. Nearly one quarter, or 24 percent, of American homes had a computer by 1994. It was around that time that the adoption of home computers began to tip, achieving a more robust pace that reached 42 percent of American homes by 1998. In 2003, according to the U.S. Census Bureau, "70 million American households, or 62%, had one or more computers."

The diffusion of the Internet in American homes was considerably more rapid than the computer. The Census Bureau's Current Population Survey began probing Americans about home Internet use in 1997. That year 18 percent of households in America reported using the Internet. At the start of the millennium, in the year 2000, four in ten households, or 40 percent, were connected to the Internet. By the close of 2001 more than 50 percent of American homes were accessing the Web. Sixty-two million households, or 55 percent, had Internet access by 2003. That was more than triple the proportion of Internet households in 1997. Nearly all households with a computer in 2003, 88 percent, had access to the Internet. Indeed, by the late 1990s the Internet was the primary motivation for purchasing a computer, as the two, in effect, became synonymous. Our lives, needless to say, have never been the same.

The generation of young people we met came of age in technology-rich households. In fact, they were the first generation of American teens to grow up with computers and the Internet literally at their fingertips. It was their presence in the household, more than any other factor, that correlated most consistently with the presence of computers in the home. In 2003, 76 percent of homes with school-age children, six to seventeen years old had a computer compared to 57 percent of homes without kids. Also, homes with school-age children were more likely than homes without them to be connected to the Internet, 67 and 57 percent, respectively.

Not surprisingly, many of the young people we talk to share stories of how the Internet has become a routine part of their everyday lives, shaping how they learn, live, play, and communicate with their peers. Many of them were introduced to computers at an early age, around nine years old. Many of their earliest memories involve computer games, the gateway experience to computers for most children. But not long after that, many of the young people we met were introduced to the Internet. As twenty-one-year-old Jonathan told me during an interview, "I can't imagine living without computers because I've never really known a world without them." Like many of his peers, Jonathan has also never known a world without an Internet that offers unprecedented access to information, entertainment content, and, most important, his close circle of friends.

The initial attraction to the online world for many young Internet users was e-mail. Twenty-year-old Allison recalls e-mailing her friends when she was ten. "At the time, "Allison said, "e-mail was the cool thing to do and it was new and a lot of fun too." Allison laughed at herself now: "I would call my two closest friends and ask them to go online and respond to my e-mail." Early in the Internet's history, researchers often considered e-mail the "killer app" because of its heavy use. Young people's new media behaviors turned a pivotal corner in 1997. That was the year AOL Instant Messenger (AIM) was introduced and became an immediate hit with teenagers.

When we asked young people to describe their first true adventures online, they easily shared vivid memories of IM and the time they spent communicating with their friends on the service. As young teens they rushed home from school to use IM. For decades, when American youth arrived home from school, they turned on their television screens. But the enthusiastic embrace of IM reversed, almost overnight, a four-decade-old habit of daily life in America. IM was a way to extend the time teens spent with their friends. The rise of the instant messaging generation was a harbinger of things to come, namely, the Internet as an emphatically social and communal space. It would take a while before the larger public began to realize what was happening, but young people were making their way to the digital world.

For the first generation of online youth, IM was one of their first truly independent experiences with the Internet—that is, time spent online alone rather than under the direct supervision of an adult authority figure like parents and teachers. It was around this point in their lives that they began going online, not because someone thought it was a good educational or novel activity but rather because they wanted to. Research suggests that the early teens, ages thirteen to fourteen, represent a digital tipping point. A 2005 Pew Internet & American Life Project report writes, "Starting junior high seems to be the moment when most teens who were not previously online get connected." In 2005 about 60 percent of the sixth graders Pew surveyed used the Web compared to 82 percent of seventh graders. Among twelfth graders, 94 percent were using the Internet. Online services like AIM were tailor-made for teenagers transitioning toward their own peer networks and greater independence from adults. At a time in their life when the world as well as their own bodies and behaviors were undergoing profound changes, adolescents were offered a chance to assert a modicum of control over their lives with IM. Later, with the rise of social-network sites like Friendster, MySpace, and Facebook, they asserted even more control over their bodies by producing and performing elaborate online identities.

Twenty-year-old Victoria believes that IM was thrilling and liberating at the same time. "IM was like the best of both worlds," she recalled. "I could do my homework, chat with my friends, and surf the Internet at the same time without getting into any trouble." IM was also a great way to get the dish on all of the latest drama in school. In Victoria's words, it became "kind of addictive." In fact, her parents, like most other adults, had no idea how much Victoria used IM. Young people's migration to digital left an indelible mark on family life. Many parents found themselves confronting new challenges regarding the impact of computers in their children's lives. A *New York Times* article on the phenomenal role of IM in young people's lives maintains that the application turned many teens into "the overconnecteds," that is, a generation of youth that became obsessed with the state of almost constant connection to their friends and social networks. Before long, Victoria, like many other teens, found herself awake and online as late as one and two o'clock in the morning on school nights. After a number of bleary-eyed mornings and suspicion that Victoria was doing more than homework on her computer, her parents established stricter rules that limited her time online.

Despite all of this, the young people who grew up in technology-rich homes were no different than the generations of youth who preceded them. Like most teens since post–World War II America, the so-called "digital natives" eagerly embraced opportunities to break away from their parents and establish their own cultural milieus, independence, and identities. It just so happened that for this and successive generations, digital technologies allowed them to branch out in some hyper-efficient and extraordinarily creative ways. The use of e-mail and IM in the 1990s established one fact about young people's online behavior that remains true today: staying connected to peers is paramount. According to one group of researchers, "IM satisfies two major needs in adolescent identity formation—maintaining individual friendships and belonging to peer groups." Young people were drawn to online platforms that facilitate opportunities to develop extremely strong, persistent, and real-time ties to their peers while also interacting with a wide range of cultural content such as pictures, music, and video.

From the very beginning of home-computer adoption, school-age children embraced the technology like no other segment in America. In fact, if you go back and look at the first academic studies of home-computer life, one theme stands out: young people enthusiastically embraced the technology. As computers began entering more and more homes in the late 1980s, many adults shied away from the technology. Children and young teens, on the other hand, gravitated to home computers. They learned, played, and, most of all, experimented with the technology. In the beginning it was mostly boys, but soon thereafter girls developed great interests in computers too.

In 1995 Robert Kraut, a Carnegie Mellon University professor, and his colleagues launched the HomeNet field trial to learn more about home Internet use. One of their initial experiments involved supplying forty-eight families in Pittsburgh with a computer and Internet service. The sample cut across a diverse spectrum of America. Fifty-seven percent of the participants were female. Roughly 25 percent had an annual family income under $35,000. A fifth of the sample, 20 percent, had never used computers before. And 24 percent were from non-white households. The Carnegie Mellon team investigated a variety of things, including where the computer was located as well as who used the computer and for what purposes and for how long. At the time of the HomeNet project, roughly 37 percent of American households had computers.

During the first four weeks of the Carnegie Mellon study, teenage males and females were logging in about eight and six hours a week online, respectively. Adult females came in at around one hour a week, whereas adult males spent less than an hour online. Teenagers, the researchers discovered, were the family gateway to the Web. In one of the first attempts to report on their findings, the Carnegie Mellon group wrote, "Teenagers' enthusiasm motivated other family members to use the Internet, and their skill helped them overcome barriers." Of all the variables that the researchers examined—including, for example, race and ethnicity, gender, education, household income, and social extroversion—none was a more powerful predictor of computer use than age. Kraut and his colleagues reported that parents had to impose time limits on their children's computer use. When it came to personal computer use in the home, young people were true early adopters. The question, of course, is why?

It turns out that teens were especially attracted to the applications that allowed them to connect with their peers from school. No matter if it was e-mail, Internet Relay Chat, or multiuser dungeons (MUDs), teens' use of the Web was principally social. Though the metaphor of the "information superhighway" dominated early public discourse about the Internet, teens realized early on the value of the Web as a social and communal tool. As I discuss in subsequent chapters, this is a resonant and consistent theme.

In the migration to the digital world the power dynamics between adults and young people were often flipped. Indeed, it is fairly common for adults to learn some of the wonders of the Internet from the innovations of young people. In one of the first research papers produced by the HomeNet Project, researchers wrote that teenagers "became the technical support gurus of the family, and would be consulted by their parents and younger siblings about technical problems and goals." Many

businesses are mixing IM into their culture to communicate more efficiently and effectively. And, after seeing how young people built vibrant
online social networks, many adults joined platforms like MySpace and
Facebook to rekindle long-lost friendships and LinkedIn to broaden
and cultivate their professional networks. Teens, many researchers discovered, led the transition to digital.

Between 2005 and 2006 the adoption of broadband took off just as many
of the older teens and young twenty-somethings that we talked to hit
the adolescent years. According to the Pew, "The availability of a broadband connection is the largest single factor that explains the intensity of
an American's Internet use." As recently as 2004, most Internet homes
in America were not broadband. But by 2006 a considerably wider
cross-section of Americans was moving to the broadband world. Between March 2005 and March 2006, broadband adoption grew 6 percent in households with annual incomes between $40,000 and $50,000.
During that same period, broadband adoption grew 121 percent for African Americans and 46 percent for English-speaking Latinos, turning
what Pew once called the "broadband elite" into a steadily evolving and
increasingly diverse "broadband public."

The 25 million Americans who picked up broadband between March
2005 and March 2006 were more than the entire broadband population
by the end of 2002. But the most revealing aspects of broadband went
far beyond growth rates, adoption curves, and demographics. The real
story is what Americans, especially younger ones, did with the faster
connection to the Internet.

Broadband quite simply changed American lives by changing what
Americans did at home. Starting around the late 1950s, leisure time in
the home was the principal, if not exclusive, domain of television. Discussing television's most fundamental impact, Robert Putnam writes,
"Nothing else in the twentieth century so rapidly and profoundly affected our leisure." Television indeed was a leisure magnet and masterful
time stealer. Putnam points out that between 1965 and 1995 Americans
gained an average of six hours a week in additional leisure time, and, he
writes, "we spent almost all six of those additional hours watching TV."
But the rapid adoption and subsequent cultural impact of broadband

challenges what has been an absolute fact about American household life for nearly half a century: the dominance of television.

Many of the earliest studies of home Internet use, academic and proprietary, found that television viewing was not adversely affected by the arrival of the computer in American households. A few studies even concluded that use of the Internet increased television use. These studies suggested that heavy Internet users were also heavy seekers of information and hence media consumption of all kinds was high among them. In a 1999 study titled "TV Viewing in Internet Households," Nielsen Media Research found that "Internet homes are lighter TV viewers but analyses of the same homes before they had Internet access revealed that they were lighter TV viewers to begin with." Nielsen concluded that there was very little evidence that the Internet directly impacts television viewing.

In retrospect, these claims are reasonable and even predictable. Before broadband, the Internet was more textual than visual. Consequently, it had more in common with print media—newspapers, magazines, and encyclopedias—than it did with television. The typical user was more likely to "read" the Internet rather than "watch" the Internet. Similarly, the typical user was more likely to consume rather than create Web-based content. In other words, the kind of experiences the Internet offered prebroadband rarely if ever rivaled television or the user-generated media so common today. By making the Internet a much more visual, dynamic, and creative experience, broadband also made it more directly competitive to television.

Our research found steady evidence that the social Web is emerging as a viable alternative to television and a preferred leisure activity among young people. Our surveys and in-depth conversations revealed that young people are increasingly likely to express a preference for the Internet over television. Among the college students we surveyed, 67 percent said that since arriving at college the amount of television they watch has decreased, compared to 11 percent who said the amount they watch has increased. Though the Internet is not solely responsible for this decline, its role cannot be overstated. Also, our research found that young people spend an average of twenty-one hours online a week compared to roughly fourteen hours watching television.

They are in fact just slightly more likely to own a computer than a television, 97 and 92 percent, respectively.

The Internet, no matter how you spin it, is a force to be reckoned with in household life. This fact alone is historic if only because television has long been the "go-to" technology for household leisure, information, and entertainment.

Signs that a broadband-powered Internet was displacing television were visible as early as 2002. That year Pew reported that 37 percent of broadband users it surveyed said they cut back on their television viewing. The differences between broadband users and those who used dial-up began to reveal what were, in reality, two different Internet users and experiences. Three things distinguished broadband users from their dial-up counterparts. First, broadband users were more likely to go online on any given day. Second, broadband users spent more time online each day. Finally, users of broadband did more online activities on any given day. Access to higher-speed connections made it twice as likely that a user would go online from home *several* times a day. In 2002 Pew wrote, "The increased daily usage translates into about three and a half hours more per week online for broadband users."

Above all else, young users of broadband are especially drawn to managing and making online content. Compared to their dial-up counterparts, the downloading activities of home broadband adopters are significantly more robust. That same 2002 study found that individuals using home broadband were five times more likely to download games and videos and three times more likely to download music. The difference in media streaming was equally pronounced, as broadband users were three times more likely to watch a video clip and five times more likely to listen to an online radio station. Pew's initial assertion of broadband adoption found "that the decrease in television viewing is most pronounced among those most active in using the Internet for entertainment purposes." It is after all the capacity for broadband Internet to deliver music, video, and games on demand that makes it such a compelling alternative to television.

In addition to devouring online content, young people are passionate creators of online content. Young broadband users thrive in the do-it-yourself environment of today's online world. By 2004 more than half of American teens reported that they created content for the Web. You name it—art, photos, blogs, personal Web pages, mash-ups, videos, game mods—and they are creating it. Pew has a name for this bunch of young and enthusiastic Internet users: "Power Creators."

If you have ever built an avatar, joined a guild in an online game, shot and uploaded a video online, crafted a digital scrapbook on Flickr, maintained a blog, or managed your own MySpace profile, you know how immersive and time consuming life on the Web is. If you have done any of these things, you also know how disengaging television seems by comparison.

Broadband did more than expand the Web's technical capabilities; it actually paved the way for profound behavioral shifts and social transformation. High-speed Internet connections laid the foundation for what a group of savvy marketers and buzz engineers in the heart of Silicon Valley would begin to sell as the most exciting development in the Internet's brief but dazzling history: Web 2.0. More than anything, the term acknowledges the Internet's constant evolution and transition from something that we mainly consume to something that we increasingly produce.

Broadband did not create radically new online activities. Online communities, identities, and fantasy worlds existed years before superfast Internet connections went mainstream. What broadband did do, however, was expand a relatively small collection of early adopting technophiles into a massive but highly differentiated public of netizens, world builders, bloggers, gamers, social networkers, and content creators.

Television's pivotal role in American households has led many to call it the "first screen." For decades, television was the screen we turned on to get our news and information. It was the screen we turned on to connect to the world or simply to pass time. However, for a growing number of young people, a new "first screen," the computer screen, has emerged. Take Rachel, a hard-working and ambitious twenty-one-year-old. In many ways, she is emblematic of young people's move to the online world.

Between her full-time load of university courses and an arduous work schedule, Rachel said that television—at least the traditional television experience—has lost its appeal. "I don't have a lot of leisure time," Rachel explained, "and so the time that I do have, I find myself spending more and more of it with my computer." For Rachel, and for many of her peers, the computer has become a primary source of information and entertainment. It has also become, according to Rachel, the dominant screen in her life. As she is eager to check her MySpace and Facebook accounts, scan her messages, reconnect with her network of online friends, and grab the day's news headlines, the computer is the first screen Rachel turns on when she arrives home. "It's my way of connecting to the people and things I care about," she said. Almost seven in ten, or 69 percent, of our young respondents agreed that they log on to the Internet as soon as they get home from work or school. In years past, television was the first and for many the only screen that was turned on after a long day at work or school.

The diminishing appeal of television in Rachel's life is primarily a result of social and lifestyle factors. In other words, it is not a deliberate or even a self-conscious rejection of television. Young people like Rachel are tuning out television not because of poor programming (though some do complain about what they perceive as a line-up of lackluster shows), but rather because of something far more ominous for the networks: the traditional television broadcast model, a nearly fifty-year-old institution, is simply not compatible with the social and mobile-media lifestyle preferred by young people. Rachel said, "It is really hard to fit live television in these days." When she elects to watch television, it is usually via her computer, mainly because it allows her to stream the programs she likes. For tech-savvy youth like Rachel, the computer screen's flexibility and anytime/anywhere playability better suits their lives. Those same features give them greater control over television and their entire media experience.

Many compare the television industry's slow response to the rise of social and mobile media to the pop music industry's response to rising new media behaviors. Even in the face of a steady drop in their prime-time audience, TV industry executives have made only cosmetic changes to their business model. Meanwhile, young people are abandoning synchronous or real-time viewing. Whereas early research and television industry executives may have underestimated the digital migration's impact on television viewing, technology thought leader Don Tapscott was among the first to comprehend what digital meant for the future of broadcast television. "TV is controlled by adults," Tapscott writes. "In contrast, children control much of their world on the Net."

In 1998, just as the network computer was spreading throughout American homes, Tapscott noted that the influence of digital on television viewership was becoming most pronounced among young viewers. The migration of the young and the digital to the Web exacerbates a challenge the television industry has long faced. It turns out that the networks aggressively target young viewers not because they watch more television than any other segment but precisely for the opposite reason: young viewers, historically, watch less television than any other segment. Young viewers are so lucrative because they are so elusive. Consequently, a premium is placed on delivering them to advertisers. Ten years after Tapscott's forecast that broadcast television as we once knew it was dead, the network bosses began, grudgingly, to admit the same thing.

Jeff Zucker, NBC-Universal CEO, was one of the first major industry figures to openly acknowledge that the search for a new broadcast model was inevitable. Speaking in December 2008, Zucker said, "So much has changed in the last ten years—even in the last five years—and none of us have taken a step back and tried to think about business from a different perspective." Zucker even recognized that the failure to develop a better understanding of the young and the digital—and what their behavior means for the future of television—could spell disaster for the networks. "We are at a critical point," he admitted, "we are in danger of becoming the music or newspaper industry or something worse."

Still, it is worth noting that television continues to be a widely used screen in young people's lives. Ninety-two percent of those we surveyed own a television set and 53 percent listed television among the top three communication technologies they use most often behind the Internet, 80 percent, and mobile phones, 75 percent. Although television remains a part of young people's media mix, it can no longer be taken for granted as the dominant or most-preferred screen in their lives. This is also true for some of the youngest members of our culture, children six years old and younger.

Even though I did not interview young children for this book, I have shared many informal conversations with parents of young children. As the parent of an inquisitive and tech-savvy seven-year-old daughter, a combination of personal and professional interests inspired my conversations with young parents. And as a member of a technology task force in a K–8th-grade school, I get the opportunity to observe young children using computers, the Web, and even SMART Boards in the classroom setting.

For the first time since the arrival of television in American homes, another screen media, the computer, has emerged in the lives of young children. For more than thirty years, parents have trusted educational media, especially television programs like *Sesame Street*, to ignite their children's interest in learning. But in the move to the online world, parents are beginning to identify more interactive sources of learning. Today, the migration to digital begins at younger and younger ages. Nowhere is this more evident than in the rising popularity of computer-mediated environments for kids.

In 2007 the growth of virtual worlds targeting kids, such as Webkinz, Club Penguin, and Stardoll, began to surge. When Hitwise.com's analyst LeeAnn Prescott compared the weeks ending April 29, 2006, and April 28, 2007, she reported that Webkinz visits were up 1,141 percent, Club Penguin visits were up 545 percent, and Stardoll visits were up 275 percent. Not only have visits to these sites increased; the average amount of time children spend on them is also worth mentioning. In her research blog, Prescott writes, "Average session times can be very long, with the average visit to GaiaOnline lasting more than 44 minutes, Neopets at 35 minutes and Stardoll at 26 minutes." In other words, kids are not only visiting these sites, they are also spending a lot of time with them. If you have ever met a child who manages a Webkinz stuffed animal or a virtual pet on Neopets, then you know how difficult it is to pull them away from the computer. One friend of mine, a mother of a seven-year-old daughter, half jokingly said Webkinz is as "addictive as crack."

So, what are the long-term implications of kids' migration to digital for broadcast television? Not good. In many American households young children are growing up spending a growing portion of their media time in front of a computer screen rather than a television screen. A 2003 study by Kaiser Family Foundation researchers Victoria Rideout, Elizabeth Vandewater, and Ellen Wartella identifies some interesting trends regarding the media behaviors of children, from infants to six-year-olds. First of all, they report that at younger and younger ages kids are asserting a growing degree of control over the media they use. For example, they request a specific television program, song, or Web site.

By ages three and four, kids show a striking level of confidence and competence with computers. Seventy percent of kids ages four to six have used a computer. About one in four in this age group uses a computer every day. The researchers write, "Those who use a computer spend an average of just over an hour at the keyboard." More than half of four- to six-year-olds has used a computer by themselves without sitting in an adult's lap, and six in ten, or 64 percent, know how to use a mouse to point and click. The very youngest members in our culture are not being socialized to use television as heavily or as often as earlier generations. In computer-mediated environments made especially for them, children are learning to build, create, and interact with each other rather than passively consume the kind of preprogrammed content that television provides. The digital migration continues unabated as children derive enormous pleasure and power from the online world.

Another way of understanding young people's move to the digital world is to consider the attitudes they hold toward television and the Internet. To gain a greater appreciation for the digital migration, consider the attitudes Americans have held about television in the past. Starting in 1959, the Roper Organization began conducting, in its words, "a series of studies to learn how television stands with the people it serves." In 1987, ten years before the Internet began moving into a growing number of American homes, the public's attitude toward television was quite good. "Positive feelings about television," Roper wrote in its 1987 report, "outnumber negatives by a wide margin." Watching television was second only to talking with friends, neighbors, and coworkers among the things American said they looked forward to doing each day. That year the Roper Report also announced a milestone in its survey of

Americans and their attitudes about the tube. The percentage of people who only mentioned television as their source of news hit 50 percent for the first time since Roper began the study. "The three words," Roper declared, "that most Americans use to describe television are *entertaining*, *informative*, and *interesting*."

In an effort to learn more about the attitudes young people hold toward television and the Internet, we asked them to respond to a series of statements that we believe serve as a measure of their disposition toward the two communication technologies and, as such, the value they place on each. What are we learning?

On almost every measure, young people regarded the Internet in more favorable terms than they do television. Take their response to the statement "The Internet is a necessity in life." Among the people we surveyed, 56 percent agreed with the statement. By contrast, 42 percent believe that television is a necessity in life. The attitudes we tracked parallel the findings by the Pew Research Center.

For more than thirty years, Pew has been surveying Americans about the products they use on a daily basis and, more precisely, if they view those products as a luxury or a necessity. In addition to other topics, Pew surveyed American's attitudes about communication technologies in late 2006. Significantly, 65 percent of Americans ages eighteen to twenty-nine say that a home computer is a necessity. By contrast, only 25 percent of Americans age sixty-five and older believe that a home computer is a necessity in today's world. Even more compelling are the generational differences regarding television. Older adults, 73 percent, are more likely than younger Americans, 53 percent, to report that television is a necessity. The fact that almost half, 47 percent, of those surveyed between the ages of eighteen and twenty-nine believe they could live without television underscores the profound cultural and generational shifts underway as young people migrate from television to the Internet.

We also asked young people to respond to this statement: "I cannot stay away from the Internet for too long." We asked the same question with the focus on television. Whereas 44 percent agreed with the statement when the Internet was the subject, 34 percent agreed when television was the subject. A majority, 52 percent, agreed that there are

Internet sites they must visit every day. When asked two similar questions about television, the findings are more mixed. First, more than half of the participants in our survey, 60 percent, disagreed that there are specific television stations they must watch every day. Television as a whole, then, is not viewed as a nightly "must-see" experience. However, 51 percent agreed that there are specific television programs that they cannot miss. What does this mean?

First, it strongly suggests that television's grip on young viewers is not nearly as compelling as the online platforms like social-network sites and online games that are a daily destination for teens and young adults. So, while the occasional "must-see" program appears on television, every night is a "must-do" night for activities offered by MySpace, Facebook, and *World of Warcraft*.

Also revealing is how young people answered the statement "The Internet relieves stress from everyday life." Over the years Americans have turned to television to relax and decompress from the rigors of work, school, and day-to-day life. This aspect of television's appeal explains why some over the years have labeled it the "plug-in drug." And it still holds true today, as 56 percent of our participants agreed that television relieves stress. But more than half, 53 percent, of our survey participants also agreed with the statement "The Internet relieves stress from everyday life," suggesting that the online world too is a way to manage our moods and escape from the rigors of daily life.

The Internet, in a relatively short period of time, has joined and in some cases surpassed television as the preferred screen in the household.

As the role of television in daily American life enters a new era, we can put its past in perspective. The truth is that Americans watched television more out of habit than any innate desire for the medium. Over the years we watched mostly because we were bored, tired, lonely, or simply because that is what Americans did for almost half a century. In short, we watched television in the past because no other communication technologies offered a more competitive option. As the Internet expands our options for leisure, entertainment, communication, and information seeking, it signals the steady erosion of television's amazing run as the dominant communication technology in American households. The march to digital is historic and marks a period when after fifty-plus years Americans began tuning out one screen, the television, and turning on another, the computer.

But what kind of society is being built as young people flee a world once dominated by television in pursuit of a world dominated by the Internet? It is a world that is at once strikingly familiar and yet remarkably foreign.

TWO

Social Media 101

What Schools Are Learning about Themselves and Young Technology Users

> We are just beginning to understand the consequences
> of MySpace, Facebook, and their impact on our schools
> and students.
>
> —Ms. Roberts, principal of Westside High School

Part of MySpace and Facebook's initial appeal among young people was the fact that even though the vibrant lives they were forming online were so strikingly public, most of their activities, communications, and identities were largely hidden from the adult world. Starting around 2005, however, the relatively veiled world young people were building online was coming to an end. Rupert Murdoch's acquisition of MySpace brought more than corporate dollars to the social Web shindig; it also brought greater public awareness and concern about the lifestyles of the young and the digital. *Time* magazine's 2006 Person of the Year tribute signaled the press's intrigue with the growing role of the social Web in everyday life. Corporate America and the popular press were not the only ones who noticed young people's move to the online world. In 2006 U.S. lawmakers turned their legislative gaze onto social-network sites.

Earlier that year a Republican pollster surveyed twenty-two Red-leaning districts in an attempt to gain a better sense of the issues that mattered most to them. The 2006 midterm elections were going to be close, and the two major political parties, Democrats and Republicans, were looking to find any edge they could in their bid to control Congress. After hearing the survey results, a small group of Republicans from the House of Representatives, calling themselves the Suburban Caucus, began molding a political strategy they believed could resonate with conservative voters. One of the issues they chose to rally around was the phenomenal rise and influence of social-network sites like MySpace and Facebook.

Convinced that this was a winning issue, members of the Suburban Caucus crafted a bill aimed squarely at social-network sites, the most popular online destination among teens. The primary aim of the Delete Online Predators Act, or DOPA, was to require any school or library that received federal E-rate discounts to block access to any Web site that "is offered by a commercial entity; permits registered users to create an on-line profile that includes detailed personal information; permits registered users to create an on-line journal and share such a journal with other users; elicits highly-personalized information from users; and enables communication among users." On July 26, 2006, DOPA was brought before the whole House floor for a vote.

One of the bill's original sponsors, Mike Fitzpatrick, a first-term Congressman from Pennsylvania, explained DOPA this way to his colleagues in the House: "Social networking sites, best known by the popular examples of MySpace, Friendster and Facebook, have literally exploded in popularity in just a few short years." Those Web sites, the Pennsylvania Republican added, "have become a haven for online sexual predators who have made these corners of the Web their own virtual hunting ground." Fitzpatrick and his fellow cosponsors maintained that DOPA would help parents protect school-age kids by shutting off the online pathways cyberpredators use to find minors. "When children leave the home and go to school or the public library and have access to social-networking sites, we have reason to be concerned," Fitzpatrick told reporters.

By 2004 a decisive majority of teens, 87 percent, used the Internet. According to the Pew Internet & American Life Project, 55 percent of teens were using social-network platforms by 2006. As the father of six children, Fitzpatrick believed that his own family biography made him especially sensitive to parental concerns about young Internet users. "People care about these issues," Fitzpatrick warned as the bill he cosponsored made its way through the legislative process. "When I go home and go to the ball fields with my wife, this is what people want to talk about, parklands and MySpace." But DOPA was more than anti-MySpace; it was, in many respects, anti-Web 2.0. And that, more than anything, galvanized a groundswell of opposition.

The most forceful criticisms of DOPA expressed deep concern about the collateral damage—the elimination of most interactive Web applications from public schools and libraries—the bill was certain to cause. DOPA was a sweeping piece of legislation that if passed in its original form would have barred students' access to more than the popular social-network sites Fitzpatrick and his cosponsors labeled "a happy hunting ground for child predators." E-mail, instant messaging, wikis, blogs, and a host of other sites that allow users to share content, ideas, and their lives with each other would have also been blocked. DOPA was striking at the very heart of what made the social Web so compelling in the eyes of many—the focus on community, collaboration, interaction, creativity, and self-expression.

Not surprisingly, social-Web enthusiasts strongly opposed DOPA. But opposition came from several corners. The American Library Association (ALA) voiced its displeasure too. Given its promotion of education and literacy, the ALA's opposition was noteworthy. The organization cited several problems with the House's technology legislation. Chief among them: the fact that it was too broad, ignored education efforts, and would likely exacerbate the digital divide by making it more difficult for the technology poor to participate in the social Web.

But the ALA's greatest fear was how the sponsors of DOPA, in the rush to vilify social media, glossed over the educational potential of the social Web. Most stunning was the bill's lack of understanding of the power and richness of social media and why it appealed to many. In its "Resolution in Support of Online Social Networks," the ALA proclaimed that it "affirms the importance of online social networks to library users of all ages for developing and using essential information literacy skills."

The cosponsors of DOPA were not only legislating against technological change; they were also legislating against social change. There was little doubt outside the halls of Congress that the architects of DOPA were out of synch with how people of all ages were incorporating the Web into their everyday and professional lives. The instant and continuous communication practices were more than a source of great pleasure for teens; they were a preferred means of interaction in the adult and professional worlds too. Similarly, the desire to connect with others through the Web was a way of life across a diverse age group. MySpace may have been the brand that teens built but by 2006, more adults than teens were creating profiles.

There was concern that DOPA's claim to protect kids might actually backfire and drive young technology users underground and beyond the reach of teachers, librarians, and other adult mentors who could help them navigate the Web more effectively and safely. And while it is true that predators exist in the world of cyberspace, it is also true that they exist in the physical world. In the latter, we teach children to beware of strangers, their environment, and suspicious behavior. DOPA threatened to take away resources that could be used to aid schools and the students they serve. A more forward-thinking approach will certainly include funding digital-media literacy programs that educate and empower young people to make smart choices in their engagement with technology.

When the sponsors brought DOPA to the floor, very little research or empirical evidence was cited to substantiate their concerns about social media. Only one study of note was mentioned during the July 26 discussion on the House floor, a 2006 Department of Justice study report titled "Online Victimization of Youth: Five Years Later." According to that study, one in seven youth are exposed to unwanted sexual solicitations. Additionally, one in three youth, the report states, are victims of unwanted exposure to sexual material. But the Department of Justice report offered no evidence related to the Internet behaviors of students at schools and libraries.

Remarkably, one of the main sources of evidence rallied in support of DOPA was a popular television show, *Dateline*'s "To Catch a Predator." After presenting what he believed was some startling data from the Department of Justice report, Fitzpatrick then turned to the *Dateline* series. "Even more startling," he announced, "has been the visual evidence offered to millions of Americans through the news outlets like *NBC Dateline*'s: 'To Catch a Predator' series." At least four other supporters of DOPA mentioned "To Catch a Predator" in their floor remarks.

"To Catch a Predator" shone a primetime spotlight on the Internet's underbelly; that place where adult predators, usually men, went looking for sexual encounters with teenage girls and boys. The show typified and even fed the public panic that made social-network sites, in the minds of some, the most dangerous of online life. Debuting in 2004, the opening episode caught eighteen men, including a priest and a college student, using the Internet to hook up with underage youth. Almost overnight, the show made online predatory behavior a national source of nonstop chatter and its host, Chris Hansen, an instant celebrity.

Members of the House of Representatives were making laws based in part on the hype and hysteria ignited by a television show that eventually came under fire for questionable tactics and ethics. One of NBC's partners was an online vigilante group called Perverted Justice. Based in Park City, Utah, members of Perverted Justice used the Internet to lure adults into salacious online chats while pretending to be teens. Observers from the legal and law enforcement fields wondered if Perverted Justice's tactics prevented pedophilia or promoted it. In a provocative profile of "To Catch a Predator," *Rolling Stone* magazine wrote that for the members of Perverted Justice, "hunting predators is both the coolest online game they've ever known and a life calling."

The premise of "To Catch a Predator"—teenagers talking to strangers online—is terribly misleading. A 2007 report by the Pew Internet & American Life Project found that 91 percent of teens interact online with people they know. Young people's migration to the digital world is not driven by a desire to meet strangers online but rather to maintain and enliven their off-line relationships. Its flaws and misrepresentations notwithstanding, "To Catch a Predator" was an overwhelming hit on the House floor as members used it to make their case against social-network sites.

When it was finally time to vote, the bill passed by an overwhelming margin in the House, 430–15. But that was as far as DOPA would get. After the 2006 November election shifted control of Congress from

Republicans to Democrats, the Web 2.0 legislation never made it to the Senate floor. DOPA may have run its course, but the debate about the steadily evolving role of the social Web in America's schools was just beginning.

Around the same time DOPA was making its way through Congress in 2006, the MacArthur Foundation was launching a $50 million initiative to learn more about kids and digital media. In 2008 the Foundation released the details from a massive study of teen online behaviors. In a white paper titled "Living and Learning with New Media: Summary of Findings from the Digital Youth Project," the collaborators write that "new media forms have altered how youth socialize and learn." The researchers who executed the study found what those who have investigated young people's online media behaviors consistently find—that young people spend most of their time online with the same people they interact with off-line. Breaking with popular opinion, the authors add, "Contrary to adult perceptions, while hanging out online, youth are picking up basic social and technical skills they need to fully participate in contemporary society." Rather than deny teenagers access to social media, educators and policy makers are advised in the report to learn more about how new technologies engage and empower young learners.

In truth a growing number of educators have been experimenting with how schools and libraries can leverage the online platforms that are a pervasive part of young people's digital media experiences. One risk taker in the education world who embraced social media early on was Beth Evans, an associate professor and electronic services specialist at the Brooklyn College Library.

In a 2006 article published in *netConnect*, Evans called on educators across the country to take the plunge into the social-media pool. Doing so, she maintains, creates an opportunity to turn formal learning environments—schools, libraries, and museums—into dynamic learning environments. "Given the popularity and reach of this powerful social network, libraries have a chance," she writes, "to be leaders on their college campuses and in the larger community by realizing the possibilities of using social-networking sites like MySpace to bring their services to the public." She should know. In March 2006 her library became one of the first in the nation to establish a presence in MySpace. Evans's experiment with social media made her a celebrity in the library world. In 2007 the *Library Journal* included her among a list of thirty individuals recognized for innovation. Eager to learn more about her experiment with MySpace, I spoke with Evans.

Evans's decision to take Brooklyn College Library into MySpace was inspired by her teenage daughter. At some point in 2006 she began to notice that her daughter, who was fifteen years old at the time, was not responding to her e-mails. One day Evans asked her daughter why she was ignoring her e-mails. "Oh, I never see them," her daughter replied. "I don't really use e-mail that much anymore. I just read what people send me on MySpace." Evans realized that her daughter was part of the next wave of young technology users, and that if her daughter was not using e-mail, there was a good chance that her daughter's friends were not using it either. She was right. A 2005 study of teens and their online communication behaviors found that they prefer applications like text and instant messaging to e-mail by a wide margin. The study writes, "When asked about which modes of communication they use *most often* when communicating with friends, online teens consistently choose IM over email." The report adds that "many teens disparage email as something for 'old people.'"

Evans described the conversation with her daughter as "one of those lightbulb moments." Around the same time, she also began to notice the rising number of students using computers in her library to frequent online destinations like Facebook and MySpace. In both her personal and professional life she was witnessing, as she writes, how young people "stay tuned to their computers, managing their stables of friends and peeking into everyone's social space." Evans told me that her library "went to social networks because that's where we felt like we could find students." It was an opportunity for the library to interact more directly with young learners. Today, the idea that a library might exist in an online community is not unusual at all. It seems as if everybody— corporations, organizations, universities, TV networks, and even major contenders for the presidency of the United States—are using social media to directly engage their constituents. But back in March 2006 it was an especially outside-the-box idea for a university library to go into MySpace. In 2006 Facebook did not permit groups to build a profile.

Evans did not have enormous expectations during Brooklyn College Library's initial journey into social-network sites. The truth is that she did not know what to expect. "I thought it would energize us and frankly be fun to do," she said.

From the very beginning, many of Evans's colleagues in New York and around the country watched with a mix of intrigue and indecision as the librarian began experimenting with social media. Some in the library community asked Evans if Brooklyn College Library's move into MySpace led to a lot of reference questions from patrons. Evans admitted to me, "I never really thought it would be a huge avenue for that and it really hasn't been." Her library uses MySpace to make announcements, unsolicited library instructions, and invitations for those who have signed up as friends. Not too long after building a MySpace profile for her library, Evans began to grasp more fully the power and potential of social media. Evans said she never could have anticipated the opportunities the move into social media has created. "When I look at it now," she told me, "I think, Wow, it makes a lot of sense that this is happening."

I asked Evans specifically what social sites have done for the Brooklyn College Library. As she explained how MySpace has "connected us with people like authors, artists, and musicians, usually people involved in culture," I could not help but notice the excitement in her voice. Evans told me that libraries and the people who work in them tend to be passive by nature.

"Usually," she reminded me, "librarians may answer a patron's reference question and then that's it. We wait until the next patron comes along with a question."

Social media, she insisted, encourages a much more assertive approach to managing a library. According to Evans, "being in MySpace is like living in the community." And being a part of a much larger community enlivens her library with a greater sense of purpose.

One example Evans shared with me illuminates how participation in social media can help educational institutions literally reimagine their role in the community. Not long after joining MySpace, Brooklyn College Library was approached by an African art collector from England who asked if he could add them as a friend. The collector also indicated that he was going to be in New York and was interested in talking about his rare collection of African art at BCL. "I realized that this was a great opportunity to do something neat for the community," Evans said. Under its more traditional tendencies the library would have never been able to identify the British art collector. In addition to posting flyers the library was able to promote the event by using MySpace to link the art collector to its community of friends. It was an educational experience in more ways than one. "Usually program planning begins well ahead of an event date," Evans confessed. "For example, if we want to do a program in the fall, we start planning several months ahead of time." Amazingly, planning for the African art program took place in about two weeks, an unprecedented achievement for the library and one made possible by the social Web.

Overall, Evans's MySpace experiment has exceeded her wildest dreams. Still, she reminded me that maintaining an active presence in the social-media world is a lot of work and one of the reasons many professional librarians choose not to do it. A day or so after we spoke by phone, Evans elaborated on this point by e-mail.

"One genuine concern librarians have expressed about maintaining a presence in a social network," she explained in her e-mail, "is that it becomes one more thing we have to do, one more place to be, and another stretching of scarce resources."

For many of her colleagues, Evans believes, "the question becomes, if we go on MySpace, what do we stop doing?" In the practical world this zero-sum approach makes sense. Still, this logic fails to consider the vitality and wonderful upside to social media.

Evans and her library had one distinct advantage that made the decision to go into social-network sites less of a risk—the older community of students they serve. Schools serving a younger population are steadfast in their near-universal decision to avoid social media like the plague. For now, fears about online predators, naïve computer users, misuse of the Internet for personal rather than educational purposes, and quite frankly a lack of information drives the decision to block sites like MySpace and Facebook on school grounds.

But even as schools deny students the opportunity to use the so-cial Web to interact with their peers, they have not been able to ignore young people's historic move online. Social-network sites, much to the chagrin of educators, are a pervasive presence in the lives of schools.

On a hot summer day I visited Ms. Roberts, principal of Westside High School, an affluent and high-performing public school. The summer break left the hallways vacant, polished, and noticeably quiet. Ms. Roberts and I sat down at a conference table in her office, which was spacious and strikingly neat. Our conversation opened with the topic of technology and the presence of computers at Westside High. "Even though we do not put a computer in the hands of every student, we are technology rich," Ms. Roberts told me. At the time of our meeting, most of the classrooms, computer labs, and two libraries at Westside were outfitted with new or recently upgraded computers. Westside was also in the early stages of going wireless. The vast majority of the students were well acquainted with computers, technology, and the Internet. "Most of the kids enjoy good access to computers and technology at home," Ms. Roberts noted.

When the subject of MySpace and Facebook arose, Ms. Roberts admitted that schools are late to the party. "We are just beginning to understand the consequences of these sites and their impact on our schools and students," she said. Like most schools around the country, Westside blocks access to popular social-network sites. Still, schools are confronting a whole new generation of challenges as they come face-to-face with the lifestyles of the young and the digital. Young people's persistent and innovative engagement with technology is forcing educators across the country to rethink a host of issues including, for instance, the role of technology in curriculum design, the personal use of technology on campus (i.e., MP3 players, cell phones), and what digital media and technology means, more generally, for the social, emotional, and educational development of school-age students.

"The technology is here and it's not going away," Ms. Roberts told me during our meeting. One of the chief challenges facing schools specifically, and society more generally, is teaching young people how to successfully navigate the digital world they are so deeply immersed in.

"If we don't teach our children about using technology responsibly, then we are failing an important part of our mission as educators," Ms. Roberts said. She likened the need for educating our youth about technology to the same kind of education we offer them in other areas of their academic and personal development.

As a member of a technology task force in a K–8th-grade school, I have participated in a number of revealing conversations with educators about young technology users. Along with addressing technology matters related to infrastructure and curriculum design, our task force has made a commitment to engaging the wider community—students, staff, and parents—about digital citizenship. That is, how to encourage students to think seriously about what it means to be a member of an online community, a citizen in the digital age.

Ms. Roberts is part of a growing chorus of educators who believe that rather than battle with kids over the technologies that they embrace, schools need to engage them in more productive ways. "Just like we teach them how to read and write, we need to teach them how to use MySpace and other digital tools more responsibly," she said. In very clear-spoken terms, the veteran principal told me, "If we leave our kids to deal with these issues on their own, then we are setting them up for failure."

Unfortunately, I learned that this is exactly what is happening to young students who grow up in neighborhoods and schools that lack the resources to provide the education and instruction that can help them make better choices as they embrace digital media.

In addition to talking with educators in technology-rich schools, I also spent some time with an incredible group of school administrators and teachers working in less affluent schools. One of my first meetings took place at Southside High School, home to students living primarily in working-class households. The school's head principal, Mr. Reed, greeted me with a warm smile and a firm handshake.

When I asked Mr. Reed about the state of computers in his school, he indicated that they were located in some of the classrooms as well as a computer lab. Southside is far from technology rich. Unlike its more affluent counterpart, Westside, there were no immediate plans to make

Southside wireless. In fact, some of the hardware is dated and that limits what teachers and students can do with the computers. At least three teachers told me that the poor state of the equipment in the classroom was actually more of a disincentive for them and their students. One teacher admitted, "If my students finish their assignment before class ends, I'll give them the option of listening to their MP3 player or getting on the computer. Most choose to listen to music because the kinds of things they want to do on the computer are not possible with the equipment in my classroom."

Nevertheless, around the country schools like Southside offer their students access to computers and the Internet. The technology tipping point for many of America's poorest schools began in 1995. That was around the time that the Clinton administration made connecting the nation's schools to the Internet, especially those in urban and rural areas, a top priority.

In his 1996 State of the Union address Clinton explained his administration's technology vision this way: "In our schools, every classroom in America must be connected to the information superhighway with computers and good software and well-trained teachers." The administration's ultimate goal was an ambitious $2 billion project that would have "every classroom and every library in the entire United States by the year 2000" connected to the Internet. Roughly two years earlier the National Telecommunications and Information Administration (NTIA) conducted what at the time was the federal government's most extensive investigation of technology and social inequality.

In its first official report, "Falling through the Net: A Survey of the 'Have Nots' in Rural and Urban America," the NTIA wrote that "in essence, information 'have nots' are disproportionately found in this country's rural areas and its central cities." The Clinton administration's determined approach to democratize access to the Web was a response to what researchers, policy wonks, and lawmakers began to describe as the "digital divide," the formation of the technology rich and technology poor.

Not everyone, however, agrees that technology is the solution to improving the educational performance and life chances of the population the NTIA calls the "information disadvantaged." As the Clinton administration moved to connect schools to the Web, a debate ensued. Critics questioned the merits of upgrading America's schools with computers and the Internet when reading and math levels were steadily plunging downward in many poor school districts. An education scholar who also happened to be a vocal critic of wiring every classroom in America called the idea "the romance with the machine," which, in his words, is "driven by this dream of a magical solution that does not exist." Despite concerns like these, by the time the Clinton administration left the White House in January 2001, the technological capabilities of public schools in the United States had been massively upgraded.

In 1995 a little more than half, or 56 percent, of white students used computers in school compared to 39 percent of African American students. The U.S. Department of Education found that 3 percent of public-school instructional rooms had Internet access in 1994. Six years later, more than three-fourths, or 77 percent, of public schools enjoyed access to the Internet. It was a dramatic turnaround, one that utterly transformed the technology environment of schoolchildren living in poor and working-class neighborhoods. For these students, schools became the most reliable opportunity to use computers and go online.

Most academic and government study's argued that barriers to the online world were principally economic and educational, which meant inevitably that race was not too far from the mix. In reality the barriers to the new media and technology landscape for African American and Latino communities were always more complex than the original "digital divide" narrative suggested. As stand-alone variables, household income, education, or even race do little to fully explain the online experiences of blacks and Latinos. However, when you add age into the equation, the complex adoption trends in black and Latino communities turn clearer. Compared to their older counterparts, young blacks and Latinos are significantly more likely to participate in the online world. They are going online from a mix of places—school, public libraries, and in a surging number of cases, home. Actually, the variation in Internet use *within* racial and ethnic groups is striking. Nowhere is this more evident than with Latinos, the fastest-growing population in America. Whereas 89 percent of college-educated Latinos go online, only 31 percent of Latinos without a high school diploma go online. In the United

States, English-dominant Latinos, 78 percent, are significantly more likely than Spanish-language Latinos, 32 percent, to go online.

Since the NTIA's initial reporting on the digital divide, a more diverse population of teens have found their way to computers and the electronic spaces made possible by the Web. Some of the most heavily recognized barriers to accessing the Web—race, household income, and parental education—are not as invincible as they once seemed. For instance, in a 2005 survey by Pew a majority of black and Latino teens, 77 and 89 percent, respectively, reported going online. Further, in households reporting less than $30,000 in annual family income, nearly three-fourths, or 73 percent, of teens go online. And in households where parents' educational attainment is a high school diploma or less, nearly three-fourths of teens, or 73 percent, said they go online. Statistical analysis of computer and Internet use by K–12 students from the U.S. Department of Education indicates that youth from poor- and modest-income homes are going online, principally from school.

The contours of the digital divide have grown more complex since 1995. As far back as the late 1990s, when policy advocates, politicians, and researchers were focusing on the access divide, technology activists called for action against what is referred to as the "participation divide." Access was only half the battle, technology activists warned. Newcomers to the Web also need technology training and education.

In retrospect, fixing the technology access problem was relatively easy as a combination of federal policies, grants, and corporate initiatives made computers and the Internet more widely available than ever before. Young people from poor- and modest-income households are accessing the Internet through schools and public libraries. Genuinely robust participation in digital publics, however, demands more than casual or occasional access to the Internet. As digital media technology evolves into a dynamic form of literacy, personal expression, and involvement in civic life, the participation gap between poor and affluent kids grows more urgent.

Since it began documenting American's online activities, the Pew Internet & American Life Project has considered a wide array of factors —race, gender, income, education, and region, just to name a few. When

it comes to understanding what people do online, no factor is more powerful than access to broadband. Individuals living in broadband homes
gain access to the Web's full potential. And yet the broadband homes,
offices, and college dormitories that facilitate substantial involvement
in computer-mediated communities are not universally available to all.
Ultimately, this means that the digital lifestyle and what it offers young
technology users—new media literacy, empowerment, social ties, and
access to social and professional networks—is not available to all. So,
even as a greater diversity of young people is online, not all digital experiences are created equal.

Ten years ago the majority of the students attending Southside High
would have been unable to participate in the social Web. But these
young people maintain relatively active lives online, especially in digital
hot spots like MySpace. In a series of focus groups with tenth graders
from Southside, we learned a lot about their engagement with social-
network sites and mobile phones. One of the unintended outcomes of
the digital divide debate is the tendency to overlook the active lives that
poor and working-class youth maintain online. When Pew conducted
its first study of teens' use of social-network sites in 2006, it found that
black and Latino youth were slightly more likely to have personal profiles and create content for the Web than their white counterparts. The
students we met from Southside are enthusiastic about the technologies that afford opportunities to express themselves and connect to each
other. In addition to the Web, they were all using mobile phones. Still,
they struggle against formidable odds. In addition to lacking access to
broadband at home, youth from low-income households, importantly,
lack access to the networks of informal education and support that make
navigating the challenges of digital citizenship more manageable.

During my meeting with Mr. Reed, I gained a clearer view of how
the digital divide has grown more complex and how social media shapes
life at his school. One particular and especially instructive case involved
an incident of *cyberbullying*, a term used to describe the mean-spirited
and reputation attacks that occur online. Cyberbullying comes in various forms and may include, for example, someone taking a private cor-
respondence—an e-mail or text message—and posting it so that others
can see. In other instances, it may involve using the Internet to spread

rumors or embarrassing pictures about someone. The Pew Internet & American Life Project found that almost one-third, or 32 percent, of teens say that they have experienced some form of online bullying. In that same survey Pew also found that teenage girls are more likely than teenage boys to be harassed online. "Older girls [15–17] in particular," Pew writes, "are more likely to report being bullied than any other age and gender group." In all of the schools I have visited, affluent or not, episodes of cyberbullying are common.

Mr. Reed explained how one Southside student came under attack from a group of young women who used the Web to spread rumors about her personal life. Though they turned out to be untrue, the rumors were extremely hurtful and caused the young victim a tremendous amount of emotional pain and stress. Hopeful that the attacks would fade away, the young woman initially said nothing about the incident to any of the adult figures in her life. But the rumors persisted. Desperate for some support, she finally approached one of her teachers and told her what was happening. After hearing about the young woman's predicament, the teacher decided to take the matter to one of the school's assistant principals. Eventually, the matter made its way to Mr. Reed.

"The young girl was in tears when she came to my office," the principal told me. "It was clear that she had been hurt and had no means of stopping the attacks." Her teacher noted that she seemed withdrawn and lonely. The young woman explained that her reputation had been ruined and that the rumors had isolated her from the other students. Even more tragic was the fact that she could not turn to anyone at home about the matter. Her grandmother was her legal guardian but, like many older Americans, was completely oblivious to the Web and her granddaughter's online activities.

The grandmother's lack of knowledge about social media meant that she was not a viable source of support for her granddaughter in this instance. After holding a meeting with the young student and her grandmother, Mr. Reed realized that the older guardian had no clue about MySpace or that her granddaughter used it. The lack of communication within their household about the use of digital technology and social media forced the teenager to confront the matter on her own.

By the time her teachers and guardian became aware of the nature and severity of the attacks, it was too late; the young woman's mental and emotional state had already been damaged.

I tell the story of the young woman's situation at Southside High in order to illuminate how youth in less affluent communities are severely underserved when it comes to gaining greater knowledge and strategies for navigating the social consequences of the social Web. Unlike their more affluent counterparts, the schools in economically challenged communities do not have the resources to offer classes or bring in an outside expert to discuss the latest developments in new media. Making matters worse is the fact that poor schools rarely if ever benefit from the social capital found in more affluent schools. Social capital in this instance refers to the intangible benefits that usually come in the form of parents who use their professional and informal networks to enrich a school's learning environment.

Let me share an example from my own personal experience that illustrates the vast advantages more affluent schools, possess compared to their less affluent counterparts. Earlier in this chapter I mentioned my involvement with a technology task force at a K–8th-grade school. Practically all of the parents at this small private school are college educated and represent a diverse community of successful professionals, entrepreneurs, and active figures in community and civic life, volunteer work, and board of directorships. The hidden benefits of self-employment, high-income occupations, professional prestige, flexible work schedules, or a stay-at-home parent allows a large percentage of parents to devote significant time and effort to the school in the form of committee and volunteer work.

In 2005 the parent council, along with the school's headmaster and communications director, noticed that the use of technology, especially among the older students, was undergoing a notable change. Translation: many of the students discovered MySpace and Facebook. Suddenly, parents of seventh and eighth graders found themselves confronted by the sticky appeal of two of the Internet's fastest growing communities. If you have ever talked with a parent whose teenage son or daughter has developed an intense love for MySpace or Facebook, you may notice the "deer in the headlights stare" some of them develop. The parent council and administrators agreed that everyone at the school—students, staff, and parents—needed to be educated about social media and teens' technology behaviors. One of the first things the school did was tap the rich informal networks parents in the school have cultivated. Doing so brought the school into contact with online innovators as well as business and thought leaders. Next, the school invited an Internet security officer from a local firm to deliver a presentation to parents and students about the Internet.

Among other things, the officer talked about the need for student's to exercise greater caution in their online activities and the kinds of personal information they post online. Within days of the security officer's presentation, parents began using e-mail to circulate various articles and reports about teen life online, creating a lively yet informal community of dialogue and learning that gave them more information and the confidence to discuss the Web with their children. The officer's comments struck a chord with many present that evening. One parent explained to me that after attending the program her fifth-grade daughter told her, "I don't think I want to get on MySpace." The mother's reply: "Good, because I don't want you to use it either." Even so, a growing number of parents at the school gave their kids consent to create a personal profile.

One mom, for example, told me that after discussing it with a few people she decided to allow her teenage daughter to use Facebook. She was afraid that denying her daughter access would drive her underground. The one caveat: the mother would be able to access her daughter's profile. This, I learned, is a common practice. A mother of a fifteen-year-old boy who recently started using Facebook told me that she occasionally looks at his page. A 2007 report by Pew writes that "41% of today's teens believe that their parents monitor them after they've gone online." This is not unusual. Back in the day, parents snooped around in their children's bedrooms looking for clues about troubling behavior they may have suspected. What are parents looking for in MySpace and Facebook? Well, a number of the parents I spoke with are not concerned about cyberpredators or the content their children are downloading. Instead, many parents see popular social sites as a way to learn more about the circle of friends their children develop.

Around the time many young people begin expressing an interest in social-network sites is also the time in their lives when they begin to desire greater autonomy from their parents. Peers, during this period, become extremely influential forces in the lives of young teens. Many parents see social sites as a way to peek into their children's personal communities in order to learn more about them and, more important, their peers. As one mom told me, "I've learned a lot about the kids my son hangs out with—their interests, activities, and even the people they hang out with."

In schools where the social capital, economic resources, and technology are relatively abundant, students, parents, and staff benefit from an environment that can nourish active dialogue and learning. In this instance the learning is mostly informal and outside the classroom, which, in all likelihood, makes it more accessible and possibly more effective. Meanwhile, students and their parents are supplied knowledge and information that can be used to make good decisions about technology usage. The crucial point is not that less affluent schools and parents do not care about technology and its consequences in the lives of the young people they teach and love. But rather that the opportunities to build this kind of informal environment of community and learning are much more challenging when the resources, social and economic, are scarce.

I specifically asked principals and teachers from less affluent schools about education programs related to technology. Four principals, for example, noted that they simply did not have the resources to bring in an outside expert. In addition, the school officials openly wondered to what extent parents from poor and working-class households had substantial enough knowledge about technology and social media to develop any real interests in attending an evening program devoted to the topic. Even though young people growing up in poor and working-class households are gaining more access to computers and the Internet, it is seldom at home and, consequently, seldom under the supervision of a parent or guardian. And parents who are not around when their children go online are less likely to engage them about the perils and possibilities of the social Web.

■ ■ ■

Another interesting way social media is impacting schools is the growing number of teachers who maintain active lives on MySpace and Facebook. Tweens and teens are not the only ones drawn to the world of
social media or the practice of using the Web to share their lives with
peers. Starting around 2006, the presence of adults in digital spaces like
MySpace and Facebook began to increase sharply. A 2006 report by
comScore Media Metrix, a digital media measurement company, notes
that "there is a misconception that social networking is the exclusive
domain of teenagers, but . . . the appeal of social networking sites is far
broader." A comScore analysis reported that by 2006 more than two-
thirds of MySpace visitors were age twenty-five or older. Once a niche
for the college crowd, Facebook opened up to anyone with an e-mail
address in 2006, setting off a rapid pace of growth. The percentage of
people using Facebook between the ages of twenty-five and thirty-four
grew by 181 percent between May 2006 and May 2007, according to
comScore.

Increasingly, teachers and students from the same school are maintaining active lives in the same online spaces. A number of teachers told
me that their students have looked them up online. This was the case
with Marvin, a young teacher I met. His students found him online and
requested to be friends with him. While Marvin interacts with some of
his students online, other teachers told me that they were uncomfortable doing this. In fact, one young teacher I spoke with reset her profile
to private after learning that a couple of eighth-grade students found
her Web page. She wanted to keep her personal life away from her students. In recent years there have been a number of news stories about
teachers getting into trouble because of pictures and other materials
they have posted online. These are issues that many of us are likely to
have to confront as digital media and communication technology makes
the line between public and private or personal and professional less
distinguishable.

In ways both subtle and palpable, digital technology is breaking
down traditional structures of power and hierarchy. Certainly, the line
that separates teachers from students can be rendered much blurrier in

the "anytime, anywhere" communication environment we live in today. Computer-mediated communication makes it more possible than ever before for a teacher and a student to carry on a conversation or develop a relationship away from school. Interaction in the form of e-mail, social-network sites, or an instant message tends to be more casual and personal than the interaction that is likely to take place at school.

One principal that I spoke with, Ms. White, indicated that this issue is on her radar. "Whenever I interview a prospective teacher for my school, I make it a point to ask them if they have a MySpace or Facebook page," she said. As our conversation shifted to the topic of social media and how it invades day-to-day life in her school, Ms. White grew more animated.

"I can't tell you how many teachers that I hire, especially among the younger ones, that have a MySpace or Facebook account." She added, "Now it's alien to me, but they do [maintain active lives online] and I know there is an appropriate way to do that."

Her greatest concern is not that teachers are active online; she realizes that is rapidly becoming the norm. What does concern her, however, is whether or not teachers understand the personal and professional risks associated with the kind of exposure made possible in Web-based communities.

Concerns about teachers participating in popular social-network sites are certainly not unique to the schools I visited. A growing number of teacher organizations and unions are beginning to inform their members about the potential perils of maintaining active lives on high-traffic social-network sites. A few days after we met, one instructor sent me an article from a teacher's employee union Web site imploring its members to avoid interacting with their students in places like MySpace and Facebook. This particular teacher and some of his colleagues are actively involved in social-community sites and were struck by the urgent warnings their union issued about online "friending."

The union offered up the following example of a post that appeared on a third-grade teacher's MySpace page as a kind of cautionary technology tale: "Why are you holding out on the [city name] girls. You need to come down here this weekend for a ladies night full of shots, getting kicked out of bars and puking, but not necessarily in that order." Allegedly, the young teacher's principal reprimanded her after learning about the post. Teachers, the union warned, should maintain the same kinds of boundaries online that are established in the classroom. Specifically, teachers are to "avoid adding students to your 'friend' or 'buddy' list, and don't post comments on students' profiles." Content posted by a teacher on a student profile, the union's Web site maintains, "could prompt allegations of misconduct, which can lead to parent grievances or negative employment actions . . . and criminal charges may be filed."

There are certainly risks involved in these kinds of interactions. But traditional boundaries and hierarchies are challenged and the possibilities for new relationships arise. Evidence of this is happening all around us as voters talk directly to candidates, newspaper readers to news editors, and students to teachers, just to name a few examples.

When I raised the issue of teachers using social media with Ms. Roberts, she told me that the real challenge is education—educating students, staff, and parents about the use of digital media. "It's a great tool, when it's used responsibly," she noted. In her view, part of the educational mission of schools today should include engaging their students about technology. "Young students," she urged, "need to know that when you send an e-mail, it's not confidential. When you post something on MySpace, you can't assume it's confidential." It is a lesson that adults and professionals must put to practice too.

"This," Ms. Roberts explained to me, "is the education piece," a part of the technology learning curve she believes that most schools are beginning to grasp. She suggested that handling this piece, though certainly not easy, is growing more manageable as a result of the educational initiatives schools like hers are actively involved in. But it is the other piece, what Ms. Roberts called "the administrative piece," that she and her fellow education leaders around the nation, she confided, "have no immediate solution for."

Much to their dismay, school administrators find themselves dealing with matters and conflicts that take place among students in the online environment. Social media is forcing school administrators to confront a relatively unprecedented dilemma: trying to determine where their disciplinary authority over students begins and ends in a teen culture fully immersed in the online world. As Ms. Roberts readily acknowl-edged when we talked, "Schools can't punish for what happens at home." But what about what happens online? Referring to social media, Ms. Roberts said, "When kids get into disagreements via MySpace or Face-book, it often spills over into the schools." Society as a whole is begin-ning to reckon with the social consequences of young people's persistent engagement with digital technology and how it is rewriting some of the most taken-for-granted rules of everyday life. In this instance the rules involve the disciplinary reach of school administrators.

Most school districts establish strict codes of conduct that individu-als involved in student organizations and teams sports are expected to honor. "Participation in extracurricular activities," one principal told me, "is not a right but a privilege that requires good conduct." But assessing the conduct of students has become much more complex in today's environment. In a steady growing number of cases around the country, young people's everyday lives, off-campus conduct, and in-discretions are appearing online. But when do the online and in most cases off-campus activities of students become a matter that requires the attention of school officials? Several of the principals I spoke with ac-knowledge that this is a growing area of concern for them and one that the education world is struggling to adequately resolve.

Ms. Roberts shared an incident with me that is likely quite famil-iar to you by now. One evening students from her school attended a party held at the home of a fellow student. Not surprisingly, in the age of phone and pocket-size digital cameras, many pictures were snapped that evening. But one set of photos eventually lead to controversy. That evening someone took pictures of members from one of the varsity team sports drinking beer, a clear violation of school rules, not to mention federal underage drinking laws. A few days later those pictures surfaced on the Internet and were brought to the attention of the team's lead sponsor. Upset by the pictures, the sponsor consulted with the princi-pal and then recommended suspending the violators for three to four weeks. But what the sponsor and principal viewed as an obvious viola-tion of school rules that required disciplinary action turned tense.

"The parents," Ms. Roberts told me, "were very resistant to any disciplining of their children by the sponsor because it was outside of the school and, in their view, beyond our authority." But the principal believed that the school district had a very firm extracurricular code of conduct, especially when underage drinking is involved. Ms. Roberts noted that even when students are off campus, they are still representatives of the school and thus subject to the codes and standards that come with participation in extracurricular activities. Despite protest from the parents, the principal approved the sponsors' recommended suspension.

Making these kinds of judgments about discipline are growing more common for school administrators. However, several of the school officials that I spoke with are uncomfortable with addressing student behavior in online environments. While Ms. Roberts readily acknowledges that the lifestyles of the young and the digital are an unavoidable aspect of her school's life, the principal maintains that she and her staff are not obsessed with students' online activities. "I want you to know," she stressed to me, "that we don't sit down and go to MySpace or Facebook. We are way too busy to do that. We handle that when it comes to our attention."

A young affable middle school teacher I met talked frankly about his students' obsession with social-network sites. "They live for MySpace," Mr. Walker told me. He taught in a school that was home to the children of working-class black and Latino families. Mr. Walker was delighted to see that his students were using the Web. Still, he was troubled by some of their online behavior. During our conversation he suggested that I see for myself the kinds of online identities his students were creating.

Instantly, I noticed some interesting patterns regarding how this community of teens used social-network sites. In nearly all of the twenty or so profiles that Mr. Walker and I looked at, the computer-mediated identities were incredibly theatrical and aspirational. Specifically, his students were living out many of their fantasies through the identities they created on the Web. Among other things, students exaggerated their incomes, occupations, and social statuses.

Much of the "identity work" in the pages Mr. Walker shared with me drew inspiration from hip-hop culture. The young men and women used popular rap tunes to establish a mood and milieu for the online identities they assume in MySpace. Music, for decades, has been a crucial aspect of the self-creation practices of young people. In addition to uploading their favorite rap songs, teens also upload pictures and videos of their favorite hip-hop performers. As we scanned the contents of MySpace profiles, Mr. Walker noted how many of his students were striving to appear older. This, it turns out, is a recurring tendency in young people's online behavior. A study published in the *Journal of Applied Developmental Psychology* found that while most teens rarely pretended to be someone else while participating in chat rooms, IM, and social sites, 86 percent of teens did pretend to be older online. Mr. Walker pointed out several instances of his students inflating the age listed on their profiles. One thirteen-year-old student, for example, listed his age as fifteen. Likewise, a fourteen-year-old pupil listed her age as sixteen. Other examples of teens trying to appear older include posing with cigarettes and alcoholic beverages.

In most instances, though, the desire for a more mature persona took on a decidedly sexualized tone, a form of identity work that worried Mr. Walker. There is nothing particularly new about teens expressing an interest in their sexuality. In fact, teens have long associated sexuality with greater independence, personal control, and a path to adulthood. Many adolescent researchers believe that teens' exploration of sexuality occurs during a period of immense physical, hormonal, social, and emotional turbulence. Not surprisingly, teens use MySpace to negotiate this period of change.

Teenagers may not own much but they develop a very early and clever sense of the most important thing they will ever own: their bodies. In MySpace, teens take great pride in their rapidly changing bodies and use them quite literally to articulate what I call the "aspirational self." The incessant desire to control and use their bodies as a source of pleasure and personal expression is a key theme in young people's journey toward greater social, emotional, and physical maturity. In the MySpace universe this is realized in spectacular fashion.

The young women express their sexuality in ways that reinforce many of the strict codes of femininity in popular media culture. They flirt with the camera and strike poses that are simultaneously provocative and submissive. The young men flirt too. Teenage boys play to the gaze of their peers and subscribe to tried and true notions of masculinity. In MySpace, for example, they perform their gender and sexual prowess through an assortment of hypermasculine poses that proudly displays bare chests, meticulously placed tattoos, and flexed muscles.

On the one hand, these are expressions of their newly found personal freedom as teens, increasingly aware of their maturing bodies and beginning to mark out territory to explore their sexual selves. On the other hand, these expressions are influenced by the meticulously packaged roles that the entertainment industries sell to young people. Typically, these roles are more limiting than liberating, imprisoning instead of empowering.

While doing research for this book yet another digital-media-based buzz word entered our vocabulary—sexting. This is a reference, primarily, to teens who use their mobile phones to send nude or partially nude photos to their boyfriends and girlfriends. Such acts are usually restricted to the more relatively private domain of mobile phones. And yet an image sent to a friend or a companion can eventually end up online or, as one sixteen-year-old girl and her family discovered, printed, distributed, and in the hands of school officials. A sexting incident in a Pennsylvania high school involving fourteen- and fifteen-year-old girls and sixteen- and seventeen-year-old boys led to charges of disseminating and possessing child pornography. While sexting generates sensational news headlines, the coverage does very little to enhance our understanding of the choices young people make in their use of technology.

First, most teens do not engage in sexting. Second, rather than prosecuting young people, we need to use these as "teachable moments" about technology, sexuality, and intimacy. Teens have long expressed fascination with their bodies and their sexuality. And over the years, the images and narratives in youth-oriented media have grown increasingly sexual. A 2005 study found that between 1998 and 2005, the number of sexual scenes on television nearly doubled. In the age of social and mobile media, teens' exploration with sexuality will likely become even more curious and adventurous but not necessarily treacherous. The ability to seek out more information and even exchange their thoughts about sexuality creates possibilities for learning how to manage and express their sexual identities in ways that are both safe and healthy.

In the end, most of what I observed was playful and not profane; indeed, exploratory rather than explicit. These teens were using MySpace to play with the sexual and gender identities that are familiar to them and their peers. This explains why many of the young women and men posted pictures of celebrity rappers, singers, and athletes alongside images of themselves. The paranoia surrounding social networking notwithstanding, these environments can actually be a safe place for young people to try out new identities with few risks. Teens use MySpace like a mirror. They try on an identity, look at themselves, and modify as they see fit.

In MySpace the behaviors of teens vividly illustrate what youth culture researchers have noted for many decades: young people build identities that are fluid, constantly in flux, and intensely performative.

Behavior that once appeared strange to many—sharing our lives online with others—is now an ordinary feature of everyday life. It is an example of how technology and, more precisely, how our engagement with technology is reshaping our day-to-day behaviors, cultural norms, and relationships. In his book *Smart Mobs: The Next Social Revolution*, author Howard Rheingold reminds us that one of the consequences of technologies that facilitate more cooperation and instant communication is the loss of privacy. I would also add the loss of control over our personal lives and the intimate data that defines who we are. Rheingold writes that "in order to cooperate with more people, I need to know more about them, and that means that they will know more about me."

Of all the issues raised by social media, none is more salient than the matter of privacy. Notions of privacy, how we define and experience it, are undergoing undeniable change in the age of what Chris Anderson, editor in chief of *Wired* magazine, calls "radical transparency." According to one survey of teens, 79 percent of respondents include pictures of themselves in their profile. Sixty-six percent say that they include pictures of their friends in their profiles. The ubiquity of digital cameras and life-sharing media highlights how we have become our own pa-

parazzi, which forces us to rethink what cultural anthropologist Erving
Goffman calls the front-stage self, the person we present to the public,
and the backstage self, the person who is much more private. In the
digital age the backstage self is more likely than ever before to be in
view, thus blurring the lines between our private and public selves. In
the online world young people are constantly being watched by their
peers. Increasingly, others are watching too.

Many of the college graduates we spoke with told us that one of the
most common questions that they are asked during an interview is, "Do
you have a Facebook or MySpace profile?" College admissions boards
admit to looking at the personal profiles of their applicants. Rightly or
wrongly, employers and admissions committees are looking at the backstage self and gaining access to young people's private selves. When I
spoke with middle and high school students I asked them one question:
"What does your profile say about you?" The truth is we all leave digital
footprints. It just so happens that for tweens and teens the footprints
they leave behind in MySpace and Facebook will likely impact their
prospects for college admissions or employment.

Young people's migration to digital has forced us all to learn quickly
about cyberbullying, the power of media literacy, digital citizenship, and
the social benefits and social costs of social media.

THREE

The Very Well Connected

Friending, Bonding, and
Community in the Digital Age

For young people, new digital technologies . . . are primary
mediators of human-to-human connections. They have created
a 24/7 network that blends the human with the technical to a
degree we haven't experienced before.

—John Palfrey and Urs Gasser, *Born Digital:*
Understanding the First Generation of Digital Natives

Social- and mobile-media technologies are facts of life today. Ninety-
three percent of the young people we survey own a computer. Even
more of them, 96 percent, own a mobile phone. Today, the ability to
be connected to others through anytime, anywhere technology expands
our sense of place, what it means to be social, and also reshapes how we
experience community. This is especially true among the set we call the
young and the digital, the world's most connected generation ever. It
does not take a PhD or a laboratory experiment to notice, for example,
teenagers' relentless devotion to being "always on," that is, constantly
connected to some kind of a screen—usually a mobile phone, iPod, or
personal computer. I witnessed this firsthand at a family event celebrating my aunt's sixtieth birthday as my fifteen-year-old cousin spent most
of the evening using her mobile phone. If you are around teens today,
the scene is a familiar one: head and eyes pointed down toward a mobile
phone and the thumbs nimbly working the touchpad. Intrigued by my
cousin's behavior, I approached her.

"Hey, what are you doing?" I asked.

"Oh, I'm texting some of my friends," she replied.

"How many people have you been in contact with this evening?"

"About five or six," she said politely.

I continued, "Are these different conversations?"

She nodded her head yes, adding, "I've probably been in about three or four different conversations the whole time I've been here."

Over the course of the evening she spent very little time talking with the people she was in the physical presence of, mostly adults and younger children. Her behavior that evening highlights one of the many questions and debates ignited by the migration to digital: are young people like my cousin more comfortable with technology than they are with humans? To be more precise, is today's communication technology making young people antisocial? An initial and even understandable answer is yes. My cousin, for example, barely seemed to notice or speak with anybody in her physical presence that evening. But probe deeper and you come away with a different answer, one that offers insight into the shifting terms and conditions of what it means to be social among the young and the digital.

Admittedly, my cousin did retreat from the physical space she was in and the family members she shared that space with. Some researchers refer to it as "absence-in-presence." In other words, my cousin was at the dinner in body but not in spirit. But the mobile phone and text messaging also expanded her sense of space, making it possible to be with her friends even though they were physically apart. This is called "presence-in-absence" in the human communication research literature. Like many in her generation, she effortlessly conducted multiple conversations via text messaging with friends who happened to be in various locations. My cousin, I concluded, was anything but antisocial. In fact, you could make a strong case that she was remarkably social, just in ways that anyone over the age of twenty-five may simply not understand or find appropriate.

Nevertheless, an outside observer might conclude that teens like my cousin have an unhealthy desire to always be connected to their mobile phones and computers. Others might cite my cousin's behavior that evening as evidence that she is not developing the interpersonal dexterity that will enable her to cultivate the personal bonds and relationships that help build and sustain communities. As plausible as they may seem, these views in the end are seriously wrongheaded.

What I came to understand is that my cousin's true interest is not the technology per se, but rather the people and the relationships the technology provides access to. The tendency to deride teen communication behaviors as antisocial or bizarre reminds me of how we have caricatured, for instance, the amount of time women talk on the phone. In his social history of the telephone and its cultural impact, sociologist Claude Fischer writes that "conversation is an important social process, serving to sustain networks and build communities." Fischer reminds us that all talk, no matter how seemingly trivial or detached, is important. If we think of my cousin's communication behaviors that night in this light, teenagers like her emerge as socially engaged beings rather than socially disengaged misfits.

Now that the social- and mobile-media lifestyles are more routine than remarkable, more everyday than occasional, increasing speculation about the behavioral and societal impact is inevitable. In a technology-rich environment that invites constant yet on-the-go contact, how do the young and the digital maintain social bonds, relationships, and community? Are notions of community and friendship changing in the digital age? Furthermore, do teenagers and young adults believe that relationships and communities mediated by computers and mobile technologies are as rewarding, desirable, and engaging as face-to-face relationships?

In addition to talking with young people about their rapidly evolving media and technology landscape, I have shared a number of extraordinary conversations with the parents of children and teens. One common refrain of concern among parents is particularly striking: the rising suspicion that children who grow up digital will become slaves to technology and, consequently, become less social, indeed, less human. The idea that technology will dehumanize us is not unique to the digital age. Addressing this very point, Howard Rheingold, longtime author and Web commentator, reminds us that "elevators and rubber tires are as implicated as computer screens in the trend toward alienation and mechanization of human life." Curiously, technological advances have a way of not only producing a sense of gain but a sense of loss too. As we embrace new innovations we oftentimes embrace unanticipated ways of living. Talk to the parents of small children and you will detect more than a bit of uncertainty in their voices about the unanticipated ways technology and the digital lifestyle are remaking the childhood experience.

What makes parents even more unsettled today is the fact that new communication technologies are being created for children as young as two and three years old. The digital lifestyle does not begin with tweens and teens; it begins with tiny tots. So, even as parents shower their kids with the latest handheld devices, next-generation gaming consoles, computers, and smartphones, they worry that they do not spend enough time in face-to-face interactions with their peers. Parents are not alone in their concerns.

In years past, social scientists expressed serious apprehension about the media content, especially violent and sexual imagery, that's exposed to young children and teenagers. And though violent and sexual themes in media continues to be a serious topic of debate, a growing amount of attention is shifting to the proliferation of screens in homes and in young people's lives. There is rising anxiety about the sheer amount of time children and teens spend with media and technology. According to a 2006 study conducted by the Kaiser Family Foundation, kids spend between six and eight-and–a-half hours a day with media. Today, playtime for many young children usually involves time with a screen. As they observe their parents' connection to mobile phones, BlackBerrys, laptops, and other electronic gadgets, many young children mimic those behaviors. We often hear, and for good reason, that young people are leading the migration to digital. But in many homes across America, parents are unwittingly teaching their kids to be digital. In the midst of the marketing and selling of the digital lifestyle, the American Academy of Pediatrics recommends that children's daily screen time be limited to one to two hours.

Researchers also point to the larger societal consequences of early and persistent engagement with new communication and screen-based technologies. A growing number of health professionals attribute rising rates of childhood obesity in America in part to the sedentary lifestyles associated with high levels of media use. In addition to concerns about the impact of technology-rich homes on the health of individuals, concerns also arise regarding the health of community life. Research on this front expresses concern that the proliferation of personalized screens and communication technologies in our lives encourages widespread abandonment of our neighbors and informal public life. The end result: the steady erosion and eventual collapse of community and any sense of connection to a wider public and the social good.

Ultimately, the core concern among parents and researchers comes down to this: is technology turning young people into an electronic herd of social recluses—that is, a generation that prefers interacting with a computer, mobile phone, or gaming screen rather than a person face-to-face? When I met with Vince, the father of a twelve-year-old boy, he expressed worry that his son's leisure habits were indeed affecting his social development. Vince told me that his son and friends regularly get together to compete against each other in their favorite leisure activity, video games, but with what he considers a peculiar twist. The boys usually play together from separate rooms or homes. That's right: they are physically apart when they play together.

Referring to his son's choice of leisure, the father explained, "I have to literally insist that he turn off his video game and go outside and play with his friends the old-fashioned way." Vince's fear that his son is missing out on a more authentic and socially rewarding play experience is not unique; many parents feel this way. At the root of this widening anxiety is the fear that our children will become socially maladjusted and even indifferent about being in the company of people. I detect, moreover, a growing worry that we may be teaching our kids to care more about a seemingly endless array of electronic gizmos—iPods, smartphones, laptops, high-powered gaming systems, online communities—than their neighbors.

Are we?

The social changes wrought by technological changes symbolize humans' inexorable yearning for new and improved ways of living that often involve altering or even abandoning more familiar ways of life. This is even truer when you consider the debates about what sociologist Claude Fischer calls "space-transcending technologies," a reference to innovations that allow us to physically move across space (think cars) or communicate across space (think telephone) more routinely and efficiently. The Internet and mobile phones are, of course, the ultimate space-transcending technologies. Anxiety about technologies that allegedly remake our world and social behaviors is not new. It is a recurring story plot in American cultural life.

In 1933 researchers Malcolm Wiley and Stuart Rice argued that space-transcending technologies like the automobile and telephone hastened the unraveling of the social fabric in American life. Addressing concerns that social contact was becoming briefer and more detached, Wiley and Rice feared a waning "of those values that inhere in more intimate, leisurely, protracted personal discussion." The two researchers believed that technology made modern life in America less intimate, less connected, and, consequently, less of a community. For instance, the automobile purportedly made it easier to move further away from family and friends. Consequently, the opportunities to do things like share meals or simply pay a neighbor a visit were greatly diminished. The telephone, similarly, made it easier to be apart and establish what one sociologist derides as "far-flung relationships" and another calls "inauthentic intimacy."

Sound familiar? It should, because these are precisely the same concerns expressed about the relentless diffusion of new communication technologies in our lives today. Among other things, the social Web and mobile phones are accused of fostering less intimate forms of contact that in effect undercut our ability to develop the ties that bind and keep us together. And so, today's new communication technologies join a long line of innovations that presumably make staying connected more difficult. Assuming for the moment that new media are transformational technologies, are they changing how we bond and build personal relationships? More precisely, are we becoming less intimate, less connected, and less interested in being in face-to-face situations with the people we call friends?

One of the central claims accompanying the rise and selling of the social Web is the notion that the Internet is not only the world's greatest tool for connecting people to information; it is also the world's greatest tool for connecting people to people. Internet evangelists commonly refer to this as the "holy grail," that is, the ultimate and most divine promise of the Web. New communication technologies also invite us to rethink our notions of space, communication, and community. Howard Rheingold was one of the first to define virtual space as social and communal space. Rheingold admits that he experienced more than a bit of apprehension prior to his involvement in online communities back in the 1980s.

"The idea of a community accessible only via my computer screen sounded cold to me at first," he writes, "but I learned quickly that people can feel passionately about e-mail and computer conferences. I've become one of them."

Since going online, Rheingold, like millions of other people, has not only met people, he has also developed a degree of affection for them. What initially seemed cold and distant became warm and intimate over time. Rheingold discovered something else that makes online communities distinct. "In traditional kinds of communities," he adds, "we are accustomed to meeting people and then getting to know them; in virtual communities, you can get to know people and then choose to meet them."

Not everyone, of course, is convinced that the Web is a legitimate sphere for experiencing community, maintaining friendships, and accumulating social capital. For some the very idea of virtual community is oxymoronic, even farcical. Putnam, for example, asserts that virtual relationships and virtual social capital are nothing if not contradictory. Those sensitive to this view typically ask: can you truly establish bonding, mutual trust, and reciprocity with someone you interact with primarily through computer-mediated communication? Their argument goes something like this.

Let's say you meet a person calling himself BoyWonder in a virtual world and discover that you both share a mutual affection for politics. The two of you begin talking and enjoying each other's company. Before you know it, you and BoyWonder are hanging out frequently, chatting and exploring the wonders of the virtual world together. Meanwhile, you begin neglecting your off-line friends, preferring instead to spend crucial parts of your evenings with BoyWonder. Before you know it, your frequent encounters with him emerge as one of the more meaningful parts of your day as you develop mutual trust and affection for each other. One day, you suddenly find yourself facing an unexpected crisis. Your car has broken down and you need help getting to work. How likely are you to call on BoyWonder for assistance? Not likely, according to critics of online relationships. As it turns out, you do not know BoyWonder's real name or place of residence. For all you know he could be a thirty-five-year-old mom role-playing her way through the virtual universe. Critics dismiss such connections as fleeting, super casual, and simply not as deep or binding as face-to-face interactions. Noting that part of their appeal is the ease at which one can come and go, some characterize online connections as "drive-by relationships."

In contrast, off-line relationships are championed as more intimate, binding, and in the end, difficult to abandon. You can easily avoid an instant message, a raiding guild in an online game, or a Facebook poke. Avoiding a classmate, neighbor, or someone who attends the same church as you, however, is a greater challenge. Putnam's question, "Are virtual relationships the real deal?" sums up a widely held view. And yet, the question is based on a premise that has drifted out of touch with how and why new communication technologies matter in most of our lives today.

The most robust criticisms of online relationships target the affiliations formed between people who do not know each other in the off-line world. The practice of "friending" strangers in the online world has never sit well with critics of computer-mediated lifestyles. While online-only relationships continue to take place, for example, via chat rooms, Internet forums, and massively multiplayer online role-playing games, they do not describe the majority of people's day-to-day engagement with the Web. For most of us the social Web has evolved, primarily, into a medium for communicating with people we know—coworkers, classmates, family, and most notably, friends. The young people we meet do not "friend" strangers online; they "friend" friends. In some respects, everyday use of the social Web parallels another technology that began reshaping our communication behaviors and social relationships roughly seventy-five years ago: the telephone.

Writing about the early role and influence of the telephone in American cultural life, Fischer says, "The telephone began as a novelty, became business's substitute for the telegraph, and then evolved into a mass product, an everyday device for handling chores and having conversations." The telephone industry's initial promotional efforts of the technology into American homes never envisioned the telephone becoming what it became—a device people used to interact socially with others. Industry men, Fischer maintains, actually tried to suppress sociable phone calls. Needless to say, those efforts failed. Over the years, the study of the social impact of the telephone has produced mixed findings.

As the telephone became more and more of a social tool, some maintain that it simultaneously decreased loneliness and face-to-face visits. In other words, it made it easy for humans to connect while also pulling them apart. But Fischer believes that the empirical evidence supporting such a claim is either weak or in the end simply nonexistent. After carefully assessing the historical data on telephony from various countries, scholars, and the telecommunications industry, Fischer reached a different conclusion. Discussing the telephone's influence in daily American life, Fischer writes, "The number of total contacts stayed constant or increased." The adoption of the telephone, Fischer argues, did not destroy people's desire to interact with each other. Fischer concludes, "Telephone use multiplies all forms of contact," adding, that because of the telephone, "it is much likelier that the total volume of social conversations increased notably. The telephone probably meant more talk of all kinds."

Thus, no matter if it was to call instead of visit, call because visiting was difficult due to physical distance, or call to arrange a face-to-face encounter, the telephone, during its earliest stages of diffusion in American homes, kept people talking and in touch with each other. And that, Fischer concludes, was a good thing.

As today's preeminent tool for staying in touch with close friends, the social Web is developing a similar role among young people today. The Internet has long been promoted as the ultimate "human network"; a communication technology that renders geography insignificant as humans form bonds, connections, and relationships across the globe. In reality, what the Web is for millions of people—a super efficient means for staying in constant contact and conversation with friends, colleagues, and acquaintances—falls short of the lofty claims articulated in more utopian narratives about the Web. This is not a critique, but rather an indisputable fact of life in the digital age. The real question is not whether or not the rise and adoption of innovations like the phone or the Web change how we interact with each other. Of course they do. Rather, the more interesting question is: do innovations in communication technology alter in any substantive or measurable way the quality of our interactions, personal relationships, and how we experience community?

The assertion that new communication technologies are making us less social, caring, and involved with others is baffling when you consider the preponderance of evidence that actually compels a substantially different question: is today's technology-rich environment making us too social, too connected, and too involved in other people's lives? In our conversations with young people, they talked candidly about the constant state of connectivity made possible by social-network sites and mobile phones. Among the young people we met, there is widespread recognition of what they call "e-stalking," that is, the degree to which their peers frequently use social-network sites to track people's lives, activities, and relationships.

And then there are social sites like Twitter, the microblogging service that allows users to follow each other throughout the day. A "tweet," the short message or status update users post to their Twitter network, answers the question, What are you doing now? in 140 characters or fewer. Journalist Clive Thompson believes that social-network tools like Twitter enable what he calls "ambient intimacy," a way to connect to people's lives, both the mundane and moving, in unprecedented ways.

A mounting body of evidence, anecdotal and empirical, strongly suggests that young people are extremely social. Discussing the slavish ways teens use instant messaging, a 2006 *New York Times* article contemplates what it calls their "seemingly obsessive need to connect." The story goes on to call this generation of teens the "overconnecteds." The empirical evidence also highlights young people's social behaviors. Seventy-five percent of American teens report using IM in a 2004 survey by Pew Internet & American Life Project. That same report found that 30 percent of teens use IM several times a day and close to half, 48 percent, use it daily. Among other things, teens use IM to check in with each other, chat, and manage and reinforce their relationships.

The rapid diffusion of cell phones adds a mobile dimension to how we manage personal relationships. In the age of constant connectivity, the young and the digital can reach out and touch each other anytime and from anywhere. And they do. No matter if they are in school, driving, or even co-present, teens habitually connect with each other through their relentless use of text messaging.

Teens and young twenty-somethings are the most inventive and incessant users of mobile phones. A 2005 report by NOP World Technology mKids found that 75 percent of teens, ages fifteen to seventeen, carry mobile phones. That same study notes that "tweens," kids ages twelve to fourteen, use cell phones too. What's more, a 2005 report from the Pew Internet & American Life Project found that 65 percent of cell-phone users ages eighteen to twenty-four use their phones to send text messages compared to 13 percent of cell users ages fifty to sixty-four. Pew describes those who have given up landlines to use cell phones exclusively, mostly the thirty-and-younger set, as a breed apart. For this growing sector, mobile phones are a constant feature in their daily lives and a multipurpose platform for communication, content creation, and life sharing.

Among young mobile-phone users, short message service (SMS) has emerged as a prominent, popular, and preferred means of constant communication and connectivity. Whereas teens in previous generations relied on letters or the telephone to connect when they were away from each other, personal computers and mobile phones have raised teen connectivity to a whole new level.

In 2007 the *Washington Post* reported on the dramatic rise of text messaging among young people. The feature article included a profile of a seventeen-year-old high school junior who sent 6,807 text messages in one month. High? Yes. Unusual? Not as much as you might think. A spokesman for Verizon Wireless told the *Washington Post* that "for a teenager to send thousands of text messages a month is not unusual." The 2008 CTIA Semi-Annual Wireless Industry Survey found midway through the year that nearly as many pictures and other multimedia messages, totaling 5.6 billion, had been sent than during the entire previous year. In the month of June 2008, Americans sent 75 billion text messages to each other. If you are counting, that translates into 2.5 billion messages per day.

From the moment teens discovered e-mail, IM, texting, and social-network sites, they understood that social media is a way to communicate and congregate.

When journalists first discovered MySpace, they immediately got one thing right by consistently describing it as a "teenage hangout." A 2005 *New York Times* article explains MySpace this way: "[It] has the personality of an online version of a teenager's bedroom, a place where the walls are papered with posters and photographs, the music is loud, and grownups are an alien species." *USA Today* wrote, "Forget the mall. Forget the movies. Forget school. Forget even AOL. If you're a teen in America today, the place to be is the social networking site MySpace." *Wired* magazine's 2006 profile characterized MySpace as "the biggest mall-cum-nightclub-cum-7-Eleven parking lot ever created." Over and over again the press portrayed MySpace as the uber-teen scene.

What drove the massive migration of teenagers to MySpace? Well, like any classic migration story, a combination of "push and pull" factors enlivens this particular plot. In many ways, social and mobile media are a natural fit and a powerful pulling force for teens. We know that the middle school period, ages thirteen to fourteen, is the tipping point for the migration of teens to the online world. Millions of teens move online because it is hip, social, viral, rebellious, and just plain fun. Teens embrace social-network sites for one other noteworthy reason—the chance to inhabit a world away from the controlling gaze of adults.

Among tweens and teens, social and mobile media embody what Ray Oldenburg calls a "third place." In his book *The Great Good Place*, Oldenburg describes third places as "a great variety of public places that host the regular, voluntary, informal, and happily anticipated gatherings of individuals beyond the realms of home and work." Examples of third places include bars, cafés, pubs, beauty salons, barbershops, and bookstores. A jovial mood and good conversation, Oldenburg explains, are prerequisite attributes for third places. We frequent third places in anticipation of finding people, conversations, and activities that make us feel good, connected, and alive. Third places supply what Oldenburg calls "spiritual tonic," or that daily pick-me-up that helps us get through the day. Writing further about the beauty and social value of third places, Oldenburg says, "The benefits of participation both delight and sustain the individual."

When it comes to pop culture capital, teens are as rich as any one else. In his classic study of Britain's mods, rockers, and punk scenes, Dick Hebdige shows how teenagers create their own subcultural style, usually through music, fashion, and language. And yet, the online world gives teens something that generations of youth have coveted but not necessarily enjoyed: a place of their own. Indeed, this migration story would not be complete without some consideration of the forces that "push" young people to online places like MySpace.

One of the main factors driving young people to online sites is the lack of places in the off-line world for them to regularly congregate and truly call their own. Despite their status as a highly desirable consumer class, there are few informal places designated for teens to gather, socialize, and bond. Faced with few places to call their own, teens have converted an array of venues—street corners, convenience stores, parking lots, arcades, and theaters—into their very own third places. For nearly three decades, one space in particular has been a favorite destination for the young to gather, flirt, socialize, and connect—the shopping mall. In recent years, however, there has been a crackdown on the use of malls as a third place among teens with the enforcement of curfews, dress codes, and what the shopping-center industry calls "parental escort policies."

The young people we met came of age just as the computer and the Internet emerged as commonplace technologies in American households. In middle school and high school they stayed up all night IMing each other. When social sites emerged they were among the first to make them a regular part of their online and leisure experience. But as teenagers grow into young adults, their lives and technology behaviors change.

How teens behave online is not a predictor of what they will do when they become young adults. In fact, as they get older and their behaviors change, the Web, once a place for gathering and hanging out with friends, becomes a very different kind of place.

■   ■   ■

While it is common to describe young people as more comfortable in front of a screen rather than a real person, we simply do not see sufficient enough evidence to substantiate the claim. Yes, many of the young "twenty-somethings" we met hit adolescence just as personal computers, instant messaging, chat rooms, and social-network sites began penetrating millions of American households. Still, they do not choose intimacy with their screens over intimacy with their peers. In fact, we see no compelling evidence that the migration to the online world has led young people to abandon their off-line lives. What's more, the evidence suggests the opposite effect: that young people are using communication technologies to facilitate face-to-face interactions across a dynamic mix of informal places.

Both our surveys and in-depth interviews confirm one widespread notion about the Internet and the social behaviors of young people while dispelling another. Yes, it is true that the social Web has become a place where young people spend a great deal of their leisure time. Even so, the social Web has not become a substitute for face-to-face interactions. Our findings are consistent with those of other researchers: young people use the Web as a tool to engage and maintain real-world friendships and connections. In other words, use of the Web is a way to fortify rather than forfeit their off-line relationships.

Eighty-four percent of the young people we surveyed do not believe that the online world is a more exciting place than the off-line world. Young people still find enormous pleasure in the opportunities to be in the actual presence of their friends. A quick note: all of our survey respondents were university students, a time in the life cycle in which the occasions to gather with friends, dorm mates, classmates, and fraternity and sorority associates are steady and frequent. And yet, young collegians also reside in an environment—the university campus—where access to the Web is continuous, and computer-mediated and mobile communication ubiquitous. Virtually every campus space that they use, from the classroom to the dorm room, offers high-speed connection to the Web. In this environment they truly are "totally wired." And yet, easy access to the Web and the virtual connections it offers has not diminished their enthusiasm for interacting with their friends the old-fashioned way, face-to-face.

Young adults do not believe that the online world, when compared to the off-line world, is a more desirable place. To put this claim in perspective, this does not mean that young people do not enjoy the opportunity to communicate and connect with each other online—they love doing this. What it means is that they do not see computer-mediated communication as more satisfying or rewarding than face-to-face time with their friends or that the online world is a more desirable place than the off-line world for hanging out with friends. Unlike teenagers, young twenty-somethings do not IM throughout the night with friends. Unlike teenagers, young twenty-somethings do not hang out online with each other.

There is widespread speculation that given how vigorously young people communicate via the Web or mobile phones, they have abandoned more traditional ways of interacting with their peers. We constantly hear that new communication technologies are reinventing how the young and the digital communicate with each other. And it is true. Computer and mobile-phone screens are certainly the dominant screen technologies in their lives. On average, young people in our survey spend about twenty-three hours a week online compared to about fifteen hours watching television. A considerable portion of that time, they told us, is devoted to social-network sites.

Our research strongly suggests that even with the massive migration online, many of the experiences that built trust, bonding, and mutual respect in the days before the rise of the Web remain vital. You may be surprised to learn, then, that the so-called digital natives thoroughly enjoy being with each other in face-to-face situations. Even the four-pack, the group of young men and enthusiastic gamers I followed for six months, spent time with their gaming buddies away from *Halo 3*, the Wii, and the many other occasions they find to game.

Four-pack member Derrick and I talked about this during one of our conversations. "On a typical day, how many hours are you online?" I asked.

"Oh, about three to four hours a day. I'm always online," Derrick told me, adding, "When I wake up in the morning I check my messages, open up my messenger file, and check my Facebook."

Though he is online several hours a day, Derrick said that he rarely hangs out with his friends on the Web because they hang out regularly with each other off-line. They often get together to play games, participate in intramural sports, and go out on the weekends. In conversation after conversation, young adults like Derrick explain that they do not feel the need to hang out online with the people they see regularly off-line. As twenty-year-old Ryan told me, "Why would I spend a lot of time hanging out with my friends through instant message or Facebook if I know I'm going to see them later in the day or that week?" This sentiment is echoed by many of the young people we encounter.

What does this mean? Well, first, that young people value their off-line relationships. A striking 84 percent of the people we collected surveys from disagreed with the statement, "Online relationships can be just as fulfilling as off-line ones." It was rare in our meetings and conversations to meet young "twenty-somethings" who invested their time and energy in relationships that only exist in the online world. Eight of ten disagreed with the statement "You can get to know a person better on the Internet than in person."

Second, few of the young people we spoke with said that they use sites like Facebook mainly to enlarge their network of friends. Another common question we asked people in our in-depth conversations goes something like this: "Would you say social-network sites mostly expand your network of friends or enable you to communicate more efficiently with people you already know?" What we often heard is that social-network sites are used, first and foremost, to communicate in more efficient ways with established friends and acquaintances. A majority of young folks believe that social media platforms like Facebook are most useful for complementing current relationships, not creating new ones. They are actually suspicious of relationships that only have an online component.

An additional common use of social-networking platforms is to send out invitations and announcements about events and occasions that bring friends and acquaintances together, face-to-face. Like many that we spoke with, Katy, a twenty-two-year-old government and art history major, sung the praises of this particular virtue of social-network sites.

"I do think that, at least on Facebook, that the event invitations are probably the most interesting and probably the most useful and most applicable to my life," Katy told us during a conversation. Once, when she and a friend were preparing to host a get-together, they stopped and reflected for a moment. According to Katy, "I remember having a conversation and saying, like, 'How did we ever organize parties without Facebook invites? Like, how did we do it?'"

Katy symbolizes a typical yet little discussed temperament in today's technology-rich world. Even though social sites and social computing are pervasive in her everyday life, Katy maintains an active life and set of relationships off-line too. As she thought some more about the conversation with her friend, Katy said, "And it was so weird to think that we would have to, like, call people . . . or even before invitations were added to Facebook, you know, just like message people." Like many others, Katy believes that Facebook makes it easier to find out about events that bring people together. Social sites are a way, quite simply, to stay informed about what is happening off-line.

This question—does the Web makes us less social or more social?—has been an enduring one in the research community. Many of the first studies maintained that time on the Web meant time spent alone. Some researchers found evidence of what is called the "displacement theory." This is the idea that time spent with the Web displaces other activities, like spending time with family and friends. One study concludes that "one simply can not be engaged with others while being engaged on the Internet." But many researchers have long argued that humans are primarily attracted to the Web for the social experiences it affords. The research of one of the foremost scholars of human social networks, Barry Wellman, supports our view that time using the Web does not encourage the abandonment of off-line relationships.

Long before the arrival of the Internet, Wellman, a sociologist by training, investigated how humans build and maintain social networks. As the Internet population grew, Wellman began to study the formation of online social networks. In the late 1990s, Wellman and Keith Hampton conducted ethnographic research in one of the first wired subdivisions in Toronto. They called the study Netville. The researchers found that residents used the Internet in ways that were primarily social and communal. In Netville, Wellman and Hampton observed that the Internet "brought neighbors together to socialize, helped them to arrange in-person gatherings—both as couples and as larger groups—facilitated the provision of aid, and enabled the easy exchange of information." Wellman believes that community is not purely geographical but rather social and supportive. His studies consistently show that the Internet does not diminish social capital—the degree to which we invest in personal relationships—but instead complements the accumulation of social capital.

As the picture our data paints grows sharper, it is clear that young twenty-somethings do not view online platforms as a third place. In fact, they consistently rejected the idea that social sites are a space to hang out with friends. This is an especially intriguing finding when you consider the fact that young people are migrating steadily to digital and computer-mediated ways of life. This aspect of the social- and mobile-media lifestyle belies many of the common concerns expressed about new technologies and their effect on young people's interest in more traditional ways of bonding and connecting. Young people use the Web to communicate with friends, express themselves, and life share. And yet, they do not perceive the Web as a regular hot spot for gathering with friends. There are, I believe, a few reasons for this. Let me focus on one in particular.

Young adults' rejection of the social Web as a third place illuminates the social and behavioral changes that accompanies the transition from teen to young adulthood. First, whereas teens, due primarily to age and the school week, face a number of restrictions on their personal mobility, young adults enjoy more personal freedoms and mobility. Second, young adults also exercise more control over their free time. No longer confined by the rigid constraints of the school week or parents, they can come and go with much more independence. Finally, and most important, they have access to a wider range of informal public places than teens—dance clubs, coffee shops, and bars, just to name a few. The problem of place for young adults, then, is not nearly as pronounced as it is for teens. In fact, young, college-educated twenty-somethings—

what demographers call the "young and the restless"—are among the
steadiest participants in informal public life, causing many cities around
the United States to actively recruit them as a desirable segment for
creating a more vital cultural scene and local economy. These and other
factors make the Web, for young twenty-somethings, less of a destination for hanging out and more of a place to life share and communicate
with friends in between the next face-to-face encounter.

There are no shortages of myths about young people's online behaviors.
One domain of steady mythmaking relates to the kinds of communities
and people the young and the digital are most likely to interact with in
the online spaces they frequent. Despite all of the hype and hysteria
about cyberpredators and the alleged thrill of meeting strangers online,
the research literature confirms that most young users of social sites do
not mingle with strangers on the Web. In reality, the people young users
of the Web interact with online are the same people they interact with
off-line.

A 2007 report from the Pew Internet & American Life Project concludes that "91% of all social networking teens say they use the sites
to stay in touch with friends they see frequently." Among our survey
of late teen and young twenty-somethings, more than half, 58 percent,
agreed with the statement "I use social networking sites to stay in touch
with friends I frequently see in person." Pew also found that among
teens, "82% use the sites to stay in touch with friends they rarely see in
person." Pew's finding among teens parallel our findings among young
adults—a decisive 87 percent of our survey participants indicated that
they use social-network sites to stay in touch with friends they rarely get
to see in person. Specifically, many of the cases we encounter involve
individuals using social sites to stay in touch with friends and acquaintances once they have moved away to attend college.

Determined to learn more about this particular use of social-
network sites and the implications for bonding and connecting in the
digital age, we held a number of in-depth conversations with young users of social-network sites. On a brisk but bright December day, Elaine,
a soft-spoken twenty-five-year-old California native, sat down to talk
with me. As we conversed about her use of new communication tech-

nologies, Elaine explained how MySpace helped her reconnect with someone who had moved to the Midwest when they were childhood friends. Growing up together in California, they were inseparable. "If I wasn't at Lynn's house, she was at mine," Elaine told me. When the two BFFs (best friends for life) first heard about Lynn's family decision to relocate, they cried. Initially, the move created physical and social distance, and then, as time moved on, emotional distance.

Years later, despite being apart, Elaine found herself occasionally thinking about her dear friend. She remembered her e-mail and decided to see if she was on MySpace. "Finding her was like finding a long-lost friend," Elaine confided in me. Since reconnecting via the Web, the two have rekindled their friendship and are constantly in touch with each other through the Web and their phones. Sipping tea, Elaine said thoughtfully, "We don't get to see each other in person these days, but I believe MySpace has made it easier for us to stay in each other's lives." Without the site, Elaine thinks the distance would have eventually eroded their ties to each other.

One of my most interesting conversations took place with Chase, one of the young men from the four-pack. Chase also uses social-network sites to stay in touch with far-away friends. A Facebook user since his high school senior year, Chase said that his engagement with the site is steadily evolving. Among other things, he uses it to communicate with friends who are both near and far. Chase described himself as horrible "when it comes to picking up the phone and calling people," admitting that he would much rather write something like, "Hey, how are you?" on a friend's wall.

"They can write you back the next hour or the next day, whenever they have the time," he said. Chase refers to these as "passive conversations" and believes that they help manage what can often be incongruent schedules among friends separated by distance, work, school, and other life-course changing developments.

Some critics argue that these impersonal conversations simply do not measure up to face-to-face talk. Relationships that are managed in this manner, the argument continues, are severely weakened because the communication is often distant, detached, and disengaged. But there's more to the online communication between Chase and his friends than meets the eye. How they use social-network sites to communicate represents much more than simple "drive-by relationships" that are here today, gone tomorrow.

Clive Thompson contends that our use of awareness tools like Facebook and Twitter reveals an interesting paradox. "Each little update—each individual bit of social information—is insignificant on its own, even supremely mundane," Thompson writes. "But taken together, over time, the little snippets coalesce into a surprisingly sophisticated portrait of your friends' and family members' lives, like thousands of dots making a pointillist painting." In other words, the short messages and mini status updates shared regularly among friends and acquaintances via social-network tools tell intriguing stories about who we are and how we are feeling. In the end these tools can create varying degrees of intimacy.

When I asked Chase whom he interacts with most while using Facebook, he quickly identified two main sets of acquaintances. "Most interactions related to events and meetings take place with people from the University," he told me. It is, in fact, quite typical for young people to use Facebook, to coordinate their social, academic, and personal calendars. In our survey, more than six out of ten, or 63 percent, said that they use social-networking sites to make plans with their friends as well as organizations they belong to. I believe this use of social-network sites has grown even more common among young collegians.

When it comes to the notes Chase posts on the walls of others—the truly personal and casual stuff—this happens most regularly with friends who live far away from him. In these instances the notes are about things that are going on in their lives, the ups and downs that can make life unpredictable, joyous, disappointing, and meaningful in both small and big ways.

"Notes," Chase told me, "are a way to stay in touch with my friends and keep track with what's happening in their lives."

As we continued to talk about the presence and influence of Facebook in his life, Chase told me that he likes the ability to have conversations that do not have to happen right away and take up large amounts of time.

"I really don't have an hour or two hours to sit down and call everyone I'm supposed to talk to," he said. "So, it's a lot easier for me to send them a quick message like, 'Hey, I remember you had a test today, how did it go?'"

The idea that these kinds of remote conversations are a routine part of everyday behavior and communication no doubt unnerves those who already view computer-mediated communication as weak at best and inferior at worst when compared to interactions that take place in person. Chase's behavior illustrates how new communication technologies enable us to keep up contact with more people, albeit in ways that are typically brief and casual rather than extensive and always deep.

After surveying eight hundred college student researchers from Michigan State University, Nicole Ellison, Charles Steinfield, and Cliff Lampe found that they use social-network sites in the ways that both Elaine and Chase discussed with us. Social-network sites, Ellison, Steinfield, and Lampe explain, are a way that young people not only grow their social capital but keep it too. The Michigan State scholars refer to this use of social-network sites as "maintained social capital." Addressing this phenomenon, Ellison and her colleagues write, "Highly engaged users are using Facebook to crystallize relationships that might otherwise remain ephemeral." They declare, "Additionally, the ability to stay in touch with these networks may offset feelings of 'friendsickness,' the distress caused by the loss of old friends."

Is there a tradeoff in terms of how we manage and maintain friendships in the digital age? Are we adding more friends (quantity) but losing some of the nuance and intimacy (quality) that makes personal relationships potentially rich and meaningful? Is the way the young and the digital form bonds and friendships changing in any fundamental way as a result of the social Web and mobile communication technologies?

Within sociology the complex web of individual relationships and group memberships people develop—also known as social networks—has been an enduring site of investigation. Sociologist Mark Granovetter's work shows that our social networks are structured into one of two categories, what he calls strong and weak ties. Strong ties may involve a relatively small group of individuals with whom we share some of the most private details of our lives with, turn to in a time of personal crisis, or vacation with. These connections usually consist of family members and intimate friends. Weak ties, on the other hand, may consist of people we know because of a mutual friend or only see occasionally yet are still regarded as a part of our social network. If strong ties are made up of friends, weak ties are made up of acquaintances. If strong ties are the connections that help you get through the day, weak ties, it is often said, are the connections that help you get ahead in life.

For all that is seemingly new about social-network sites, they do not appear to be radically altering the personal bonds and connections that young people make. Like the generations before them, young people today still maintain strong and weak ties. In short, they develop different kinds of personal relationships that involve different levels of social and emotional investment. This was true before the Web and, as we are learning, holds true now that the Web is a widely used networking tool. What does appear to be changing, however, is how the young and the digital maintain and reinforce their strong and weak ties via the social Web and mobile phones.

As time passes we will undoubtedly gain a fuller understanding of the social implications of the digital lifestyle and what effects if any it has on how we form social bonds. For now, here is what our research tells us. Young people's migration online pivots primarily around two interests: strengthening strong personal ties that originate off-line while also keeping alive their weak ties to acquaintances. Social-network sites are used primarily to engage a close circle of friends and acquaintances. This is not too surprising when you consider the fact that their main motivation for using new communication technologies is geared toward maintaining personal connections (getting through the day) rather than professional connections (getting ahead in life). One can imagine that as they get older that many young people will expand their use of social-network sites, using them as a tool to get ahead by taking advantage of what Granovetter calls "the strength of weak ties." Anecdotally, I see evidence that this latter use of social-network sites is already happening among students who graduate from college.

Myths and stereotypes aside, young people's use of social-network sites have always been in a state of constant evolution. Though MySpace and Facebook steal the headlines, teens and young twenty-somethings' participation in online communities are driven by a number of factors.

Fifty-nine percent of the respondents in our survey of college-aged persons subscribe to two or more sites. A 2007 study by comScore found that teens who visit both MySpace and Facebook tend to spend "more time at each site than those who visit just one or the other." In addition to participating in different online communities, these young people create different online identities. Whereas participation in some sites may be for bonding, life sharing, and getting through the day, participation in others may be for emphasizing professional credentials, expertise, and getting ahead in life. Managing these different selves, interests, and networks is certainly not unique to the digital lifestyle. Still, the evolving technology landscape makes doing so more dynamic, explicit, and manageable.

Much like the transition from high school to college, the transition from college to a professional career marks yet another shift in the use of social-network sites. More and more college students told us that when they graduate they will stop using Facebook, characterizing the platform as suitable for college but not for a working professional. Others indicate that they will become much more selective about the personal data they share and with whom they share it with. After graduating, some elect to join communities like LinkedIn—a platform used primarily for growing professional rather than strictly social networks.

Whatever the trade-offs—more casual, shorter, and distant conversation—the use of social and mobile media should not be interpreted as inherently adversarial to how humans bond, build community, and cultivate social networks. The long-term net effect may be that computer-mediated and mobile-based communication complements and, in some cases, even invigorates both the strong and weak ties humans build. Researchers like Wellman believe that the Web, more often than not, complements our connections to others. The greatest benefactor in the use of social-network sites may be the weak ties that young people are able to maintain. Researchers studying social-network sites note that platforms like Facebook make keeping weak ties alive relatively cheap and easy. As Chase said, preserving a connection to an acquaintance does not take much in terms of time, by way of a quick wall post, or money—there's no need to call long distance. Social-network sites may even transform some weak ties into stronger ties, thus deepening our pool of intimate friends, people we rely on to get through the day or a tough period in our lives. Virtual-world social capital in this instance translates into physical-world social capital.

During one of our conversations, Chase and I talked candidly about his experience with Facebook. "How does Facebook influence your relationships and connections to others?" I asked.

After pausing for a few seconds, Chase said: "[Facebook has] probably made a lot of the relationships with people from high school a lot more stronger than I think they would be otherwise." He had entered school thinking that each successive year of college would lead to less and less contact with his friends from high school. Eventually he figured they would drift apart.

"That was definitely my mom and dad's experience," Chase told me.

But the social Web enables a very different kind of experience for Chase and his generation. The geographical distance caused by moving away to school does not necessarily translate into social or emotional distance. Out of sight does not mean out of mind in the digital age.

Reflecting on his own personal involvement with social-network sites, Chase said, "I've definitely noticed that my group of friends that I knew from high school is still quite large just because Facebook provides such a nice way of keeping in touch more than say, like, having to write a letter or trying to keep up with everyone by phone."

What Chase described as passive conversations, Australian researchers Frank Vetere, Steve Howard, and Martin R. Gibbs call "phatic interactions." According to the researchers, phatic exchanges do not inform; rather, they sustain sociability between people. For example, when Chase posts a note on the Facebook wall of a far-away friend simply to say, "Hey, how are you doing?" he is not providing them any information. Instead, he is maintaining a social connection with them. Of course, critics of short messages may describe the exchange as idle, impassive, or even meaningless chat. In short, illustration of a relationship that is distant, emotionally detached, and superficial. Vetere, Howard, and Gibbs, however, believe something far more substantive happens through exchanges like these. Phatic interactions, they claim, endeavor "to keep channels of communication open and to maintain the physical, psychological, or social contact." The social and community media practices that take place among friends who are far away from each other are one way to stay close and engaged despite geographical distance.

A regular criticism of the constant yet brief exchanges through social-network sites and mobile phones is that they encourage interaction that is quick and easy. Many of these interactions resemble the phatic exchanges described by Vetere, Howard, and Gibbs. Phatic exchanges may be brief, but they can also be sincere. During our conversation, Chase told me it is not unusual for him to send a quick wall note to a distant friend that says something like this: "Hi, I know you had a test today, how did it go?" This exchange is definitely quick, but it is also genuine. In other words, it is not always necessarily what is being said or even the duration of the communication but rather that something is said at all that can reflect a sense of connection and affection between people. This point is illuminated in two field-based studies of people's phatic exchanges conducted by Vetere, Howard, and Gibbs.

Discussing their research, they write that "the facility to idly chat . . . with someone you care for was a valuable expression of the care they shared for each other." The researchers conclude that the substance of the communication was not always important among their respondents. "It was the reassurance that they were connected, that a channel of communication was available to them, and that this somehow strengthened and nurtured the relationship." The authors contend that exchanges that may appear as trivial to outsiders are in fact "laden with emotional significance."

Phatic interactions, according to Vetere, Howard, and Gibbs, "maintain and strengthen existing relationships in order to facilitate further communication." As crucial as this point is, it is generally glossed over by critics and casual observers of the communication behaviors common among the young and the digital. In our research we found ample evidence that the brief exchanges between friends online or via mobile phones are seldom their only existing form of interaction. It is widely recognized that teens and young adults use mobile telephony, texting, IMing, and social-network sites to maintain what researchers James E. Katz and Mark Aakhus call "perpetual contact." What is often not recognized, however, is the fact that a post on a Facebook wall or a text message in the middle of the day translates into more than an impersonal greeting or passive nod—it often leads to a face-to-face encounter. What seems casual, cold, and impersonal—computer-mediated communication—has allowed Chase to maintain a relatively steady, warm, and personal connection to his high school classmates.

"Because of Facebook," I asked, "are you more likely to visit or spend time with your high school friends when you go back home?"

Before I could even finish the question, Chase answered, "Oh, definitely." This particular conversation with Chase could not have come at a better time: just before the end of the fall semester, and as he was preparing to go back home for the winter break.

"Like even right now," he added, "my friends and I have exchanged notes tossing around a list of ideas about what we want to do together when we get home from break."

He compared the ongoing conversations to an open community, one that gives him and his friends a better opportunity to brainstorm together since trying to call and coordinate everybody's calendars by phone would be an almost impossible task. This kind of coordination and the prospects for maintaining friendly ties despite physical distance is made more manageable by a social computing platform like Facebook. Chase agrees wholeheartedly.

"If I didn't have Facebook, I'd probably, like, talk to three people from high school," he said.

Because of Facebook, he was talking to about four times that many people. But Chase, his friends, and millions like them were doing more than simply passing casual notes online. In their own way they were investing in each other and the ties that bound them together. In their own way, they were keeping relationships alive, accumulating social capital, and building community. Despite physical distance, they were staying connected, socially and emotionally.

I began this chapter recalling my cousin's constant engagement with her phone during a family gathering. Critics look at behavior like hers and see a generation of social recluses more interested in the vast inventory of consumer electronics that they steadily accumulate rather than the people and communities around them. Understandably, the sheer growth of exchanges via social-network sites and small messaging among the young and the digital makes it easy to get caught up in the fact that they are "always on"—that is, constantly connected to a screen, a network, or some other gadget. However, in a world marked by social computing and mobile messaging, we have to make a distinction between how young people connect (the technical) and why they connect (the emotional). Certainly, my cousin's ability to text superfast or with several people simultaneously is an acquired technical skill that reflects her ease with technology. But her desire to reach out and interact, no matter how brief or seemingly detached, can also be an expression of intimacy. In short, a way to say to her friends, "I care."

Their intense connections to technology notwithstanding, today's youth are not transforming into social recluses or heartless machines. Many of the young people we met maintain a strong and healthy interest in the people around them. Our research suggests that when it comes to the way young technology users form bonds, accumulate social capital, and experience community, the social Web's influence is more evolutionary than revolutionary. Meanwhile, the social Web does not appear to be radically altering the kinds of relationships young people form but rather how they constantly and conveniently use communication technologies to manage their off-line relationships.

The young and the digital are developing distinct notions of what it means to be social and engaged with their friends. Young people simply do not believe that daily participation in the online world, posting a quick message on a friend's Facebook wall, or text messaging represent threats to community. In fact, they consistently view these and other new media behaviors as social and communal experiences.

Yes, young people are extraordinarily committed to using technology. This commitment to technology, however, is driven primarily by their commitment to each other and a desire to stay connected to acquaintances and close friends alike.

FOUR

Digital Gates

How Race and Class Distinctions Are Shaping the Digital World

> I used to have MySpace but got rid of it because it felt too open. You feel safer with a thing like Facebook. . . . It doesn't feel as if you're vulnerable to the outside creepy world. It's just your friends.
>
> —Doug, twenty-one-year-old college student

In the summer of 2007, blogger danah boyd posted an informal essay titled "Viewing American Class Divisions through Facebook and My-Space." Based primarily on her observations of MySpace and Facebook profiles, boyd ponders how class antagonisms influence young people's use of social-network sites. By her own admission, boyd was uncomfortable with the argument and the sociological vocabulary she was in search of to articulate her main thesis: that the class divisions that shape American cultural life off-line are clearly discernible in the communities that form online in MySpace and Facebook. "Americans aren't so good at talking about class," she writes, adding, "it's uncomfortable, and to top it off, we don't have the language for marking class in a meaningful way." She is right, partially.

Sustaining a serious public conversation about the class cleavages in American life is a constant challenge, but not for the reason usually cited—that Americans rarely if ever think in terms of class. The truth is nearly every facet of our daily lives—the clothes we wear, the foods we eat, the schools we attend, the neighborhoods we live in, and the company we keep—bears the visible marks of social class and the ever-

deepening cleavages between the economically mobile and the economically vulnerable.

"Facebook kids," the blogger writes, "come from families who emphasize education and going to college." Users of Facebook, boyd asserts, tend to be white and come, more often than not, from a world of
middle-class comfort. Drawing from some of the more familiar social
cliques among young people, boyd equates the "preps" and the "jocks"
with Facebook. MySpace kids, in contrast, come from the other side of
the cultural divide. According to boyd, they are the "kids whose parents
didn't go to college, who are expected to get a job when they finish high
school." Latino, black, and youth from working-class and immigrant
households, she maintains, are more likely to be users of MySpace.

In the end, boyd's essay is consistent with a concept—the digital
divide—that gained momentum as far back as the late 1990s as academic
and policy-oriented researchers began to ponder how social inequalities
impact engagement with the Internet. Some of the most vigorous champions of all things digital—the social Web, blogs, wikis, virtual worlds,
user-generated content, and social-network sites—can be intolerant of
disapproving analysis. Truth be told, technology enthusiasts pay only
scant attention to matters of social inequality. After access to computers and the Internet were widened significantly, the continued gap between the technology rich and the technology poor quickly receded to
the background.

Race is a kind of "inconvenient truth" for evangelists of the social
Web. Early in the Web's history, the anonymity of computer-mediated
communication suggested to many commentators that longstanding
spheres of social division, discord, and discrimination—most notably
race and gender—would be rendered meaningless in the digital world.
It was that logic that made the *New Yorker* cartoon about the dog and
the Internet so famous. The cartoon implies that if being a dog on the
Internet does not matter, certainly being black, Latino, or female would
not matter either. Despite the utopian view that the Web provides a
place and a way to escape the social burdens and divisions of the off-line
world, this has never been true. All of the optimism notwithstanding,
the digital world has never existed in a bubble, insulated from the social

tensions and economic inequalities that are integral to the making and remaking of the social world. Life online has always been intricately though never predictably connected to life off-line. Social inequalities still matter in the physical world. And as we are learning, they also matter in the virtual world. Nowhere is this clearer than in the rise and use of social-network sites.

Right around the time that boyd wrote her essay, my research assistant and I were assessing the data from the surveys and interviews we were collecting. Earlier in this book I explain that the use of social-network sites is the premiere online activity among young people between the ages of fifteen and twenty-four. Along with noticing how pervasive the use of social sites is among college students, we detected something else: a decisive preference for Facebook over MySpace among college students. When we asked college students, "Which social-network site do you visit MOST OFTEN?"—among white students, more than eight out of ten, or 84 percent, preferred Facebook. By contrast, 66 percent of those who identified as Latino preferred Facebook. In our survey Latino students were more likely to name MySpace as their preferred site.

What started out in 2005 and 2006 as a steady move to Facebook among American college students has become, by the time of this writing in 2009, a massive migration and cultural rite of passage. As twenty-two-year-old Sara told us, "In college you are almost expected to use Facebook." Though many of the young college students we spoke with around this time, in 2007 and 2008, began using MySpace before Facebook, they had either deleted their MySpace profile or seldom bothered to use what at the time was the world's most populated social-network site. Within months of its debut, MySpace leaped ahead of Friendster, one of the first online social-network sites, to attract a large concentration of American youth. Soon after its launch in 2004, Facebook replaced MySpace as the new digital destination for the college set. By 2007 high school students bound for college were also showing a stronger preference for Facebook.

While identifying emergent themes and trends from our survey, we noticed that the findings from a separate study conducted around the same time, in early 2007, parallel some of the results from our study. After surveying 1,060 students, ages eighteen and nineteen, Eszter Hargittai, a professor of communication studies, found that a majority of the students, four out of five, used Facebook. About one-third of the sample used MySpace frequently. But when Hargittai broke her data down by gender, race, and class, a number of interesting results surfaced. Similar to many studies, and as I note above, Hargittai found that women use social-network sites more frequently than men. But Hargittai's most interesting findings revolve around the racial and class differences her data tracked.

A majority of white students, 83 percent, preferred Facebook, whereas a little more than half, 57 percent, reported using MySpace. Eighty percent of the African American students used Facebook, and about 58 percent used MySpace. Hargittai reports some significant differences among students of Latino and Asian origins. "Hispanic students," she writes, "are significantly less likely to use Facebook (60% compared to 75% or more for other groups), whereas they are much more likely than others to use MySpace (73% among Hispanic students compared to 58% or less among all others)." Students of Asian origins showed clear differences too. Whereas 84 percent of the Asian students in Hargittai's sample used Facebook, 39 percent reported using MySpace.

Hargittai's study also shows pronounced differences across class, which she measures by parents' level of education. High levels of education attainment are often associated with higher levels of employment and income. In fact, you can tell a lot about a family's habits, activities, and lifestyle based on parental education. Past studies of children and teens use of television and video games, for instance, show a strong correlation with parental education. Children growing up in low-education households tend to watch more television and play more video games than children in households with highly educated parents. Hargittai's study finds similar results. "The most pronounced finding," Hargittai writes, "is that students whose parents have less than a high school degree are significantly less likely to be on Facebook and are significantly more likely to be MySpace users." A close look at her results reveals that the more schooling parents attain, the less likely their children are to use MySpace.

Both the survey I led and the one conducted by Hargittai confirm that something truly interesting is happening with race, class, and education as it relates to young people's engagement with social-network sites. But neither study answers the all-important question: why does racial identification appear to influence which sites students prefer?

Fortunately, we were complementing our surveys with in-depth conversations, going out into the digital trenches to talk directly with young people about their use of social-network sites.

What we learned is quite revealing.

Right away, the interviews illuminated the constantly evolving ways teens and young twenty-somethings use the social Web. Talk to them and you quickly learn that they harbor intense views, both favorable and unfavorable, toward social-network sites. Some young people are incredibly passionate about MySpace. "I use it all of the time," twenty-six-year-old Avani told us. "It's fun, exciting, and easy to meet people. I think people interact more on MySpace [than Facebook]." Loyalty to Facebook is just as strong. Frances, a twenty-two-year-old communication major, said, "It's a much simpler site to use." With assurance, Jonathan declared, "Facebook is for people who already have friends, whereas MySpace is for people who are looking for friends."

In all of our in-depth conversations we asked each person to use adjectives that, in their view, best describe MySpace and Facebook. Over the course of more than two hundred conversations with white college students, we heard all kinds of words. The preference for Facebook is undeniable. In Table 1 I list the adjectives that they use most often to describe MySpace and Facebook. Notice anything? The language they used to characterize MySpace is strikingly hostile. Words like creepy, crowded, uneducated, and fake reveal a considerable degree of bad feeling toward the MySpace site and community. By comparison, they maintain a largely favorable view of Facebook, consistently describing the platform as trustworthy, selective, educated, and authentic. "Addictive" was another common word used to describe Facebook. Along with the adjectives, young people offered a variety of stories that explain in colorful detail how they make sense of the digital media landscape and, more specifically, the two most popular social-network sites in the United States.

After analyzing the in-depth conversations, we drilled the preference among young white collegians for Facebook over MySpace down to two main factors—aesthetics and demographics. Aesthetics refers to the look, style, and manner in which personal profiles are designed and presented. The second factor, demographics, refers to the individuals and communities that tend to use both sites. Aesthetics point to the system features of social sites, while demographics alludes to system users. Together, both factors illuminate the sharp and powerful differences race and class make in the online communities young people participate in.

TABLE 1   Adjectives college students use to
describe MySpace and Facebook

MYSPACE	FACEBOOK
Crowded	Selective
Trashy	Clean
Creepy	Trustworthy
Busy	Simple
General Public	College
Uneducated	Educated
Fake	Authentic
Open	Private
Immature	Mature
Predators	Stalker-friendly
Crazy	Addictive

■ ■ ■

Visit MySpace and Facebook and the first thing you will likely notice is that the design, look, and feel of the personal profiles on each site are worlds apart. MySpace's system features encourage customization and personalization, a kind of digital wild style. The color, design, and mood of MySpace pages vary significantly. Twenty-six-year-old Avani said,

"With MySpace there's a lot you can do with HTML." Facebook profiles, on the other hand, maintain a relatively standard look. Compared
to MySpace, the tone and style of Facebook seems antistyle. From the
very beginning of its launch, MySpace carefully cultivated a demeanor
that is personal, whimsical, and, at times, oppositional. Facebook, on
the other hand, has maintained a relatively stable and uniform presentation even as it expands the scope of its services. Though the content—
pictures, wall posts, and use of applications—may vary on Facebook, the
presentation of it all does not. These two contrasting styles engender
strong views from users of social-network sites, especially those in college who believe Facebook is the superior platform.

Gerry, a nineteen-year-old sophomore, told us that "Facebook just
looks cleaner. Not now, with all the new applications and stuff, but it
still looks cleaner than MySpace and a lot more organized, as opposed
to MySpace . . . with all the background and fonts, and things." In his
words, MySpace "feels very cluttered and kind of schizophrenic to look
at a page . . . it really makes your head spin." Sarah, a twenty-one-year-
old communication studies major, agreed. "Oh, MySpace is horrible. It
takes forever to download a MySpace page and you never know if you
are looking at a real person or not."

Throughout our conversations with them, college students repeatedly expressed their dislike with the often overzealous design of My-
Space profiles and the time it takes to download them. MySpace, said
twenty-one-year-old Matthew, reminds him of the "dark, dark days of
the Internet." When asked to elaborate he said, "I don't like the fact that
the designs of MySpace pages are for the most part dreadful. They re-
mind me of back when I was seven and eight and people had just learned
how to create Web pages. And they had flashing texts and bubbles."

Likewise, twenty-two-year-old Brandon expressed irritation with
the customizable features of MySpace.

"The big difference I suppose is HTML. There's no HTML writing in Facebook as opposed to MySpace," Brandon observed. Like many
other college students, he believes the ability to write in MySpace un-
dermines the quality of the user experience. "I think it makes Facebook
so much better in the sense that you're never being plagued by someone
else's bad code."

Initially, the dissatisfaction with the customized profiles on MySpace caught me by surprise. A hallmark feature of the social Web is the ability to not only read Web-based content, but write content too. In the age of do-it-yourself (DIY) media, the fact that we are both consumers and creators of content redefines the rules of media engagement by redefining the rules of media production and consumption. And yet, it turns out that the customization and personalization of MySpace profiles through creative layouts, music, video, and graphics is a major source of annoyance and cultural friction for many college students.

Nineteen-year-old Shelby was not impressed with MySpace, a social site she believes is filled with phony names, phony profiles, and in her words, "glittery, gaudy-as-shit layouts." The repeated characterization of MySpace as "trashy," "messy," "busy," and "gaudy" unveils a widespread belief among young collegians that the profiles crafted on the platform are unrefined, unsophisticated, and unappealing. All of this is in sharp contrast to the generally glowing praise showered on Facebook profiles, which, according to twenty-four-year-old Kevin, "is much better organized and easy to use." Another young woman described Facebook as "pretty, simple, and classy."

Beneath the preference for the more uniform interface of Facebook lies a more complex tale about the influence of race, class, and geography in the digital world. The triumph of Facebook over MySpace across Campus USA is not purely about aesthetic judgments or the desire for a simple and easily navigable platform. Matters of taste, French sociologist Pierre Bourdieu reminds us, do not develop in a vacuum but rather in relation to people's social class position. Bourdieu, widely recognized as a pioneering thinker and theorist in the "sociology of culture," used the terrain of culture or, more precisely, what we do in our everyday lives, to examine expressions of social inequality. Bourdieu made a career of studying what he calls the "distinctions"—tastes, lifestyle, manners, and values—members of the middle class diligently practice in order to maintain clear boundaries between themselves and the classes they view as less cultured, sophisticated, and desirable. Sociologists following in the tradition of Bourdieu refer to these practices as boundary-maintenance work.

Bourdieu carefully illustrates how the accumulation of middle-class cultural capital—education, and a taste for the high arts and the other presumed finer things of life—does more than serve the psyche of the bourgeois classes; it also enables them to reinforce their position of privilege. Many of the distinctions college students make in relation to social-network sites are not merely about taste; they are also about the preservation of social status and privilege.

As we probed deeper into the use of social-network sites, it became increasingly clear that young Facebookers' abandonment of MySpace is not simply about avoiding "bad code"; it is also about avoiding "bad people."

In addition to developing a strong dislike for the aesthetics that define MySpace, many college students expressed an equal degree of disdain for the demographics of MySpace or, more precisely, the kinds of individuals they believe populate the community. The frequent characterization of MySpace as "trashy" and "uneducated" underscores the widespread belief among young collegians that MySpace is used chiefly by a community of digital undesirables—black, Latino, and angst-ridden teenagers—people they consistently describe as "creepy."

Twenty-two-year-old Tanner got straight to the point. "When I get on, or think of MySpace, I think of profiles that are trashy and just have garbage all over them," he said. Tanner also believes that there are class differences among those using social-network sites. "I think MySpace is the poor person's Facebook." Vanessa, a twenty-five-year-old Facebook user, also linked class to the use of social sites. She simply did not understand people who refused to use her favorite site. "People who are not on Facebook are kind of weird," she contended. Vanessa offered a sociological explanation: "It might also have to do with their socioeconomic status. Maybe they don't have a computer or can't afford the Internet."

Both Will, twenty-one, and Elissa, twenty-two, characterized MySpace as "ghetto." When we asked them to explain, Will said that while most of the people he sees on MySpace are his age, they are not in school. In his view, they are people who "edit their profile all day long and use glitter and graphics that just make everything look tacky." Will clearly associates MySpace with people who can spend more time updating their profiles because they do not attend college or hold a desirable job. Elisa said that MySpace is "ghetto" because she gets friend requests from guys that look like they are "some sort of wannabe thug with all of their chains, or their profile picture is of them posted up in front of their low-rider or something." Nineteen-year-old Cheryl shared a similar account to explain her distaste for MySpace.

"Once my friend was approached online by a guy who looked like he just came from prison and had crawled under a fence." Clearly irritated by this, she exclaimed, "We would never invite someone like that into our network, that's not what Facebook is about."

Race is not usually evoked so explicitly in our conversations with young people about social-network sites. Concerns like these ordinarily come out through more oblique statements and points of view.

Our encounter with nineteen-year-old Amy is more typical of the young college students we met. She was thirteen when she received her first e-mail account. Referring to her current Internet usage, Amy said, "I spend about five hours a day online." That time is filled with various things, including school-related work, watching music videos, and, most of all, viewing her friends' Facebook profiles. Her initial crossing into the world of social-network sites began with profiles on MySpace and Facebook. She quickly decided that Facebook offers a far superior experience. "I had MySpace for about two months." When asked how often she visits Facebook, Amy said, "It's a routine . . . many times a day." Because she has a prepaid phone account, Facebook saves her money. "I know my friends are going to check it," she said confidently.

Reflecting back on the brief period when she used both sites, Amy said, "MySpace was creepy, weird, and complicated." The college sophomore elaborated. "Creepy and weird because you can just make up a MySpace page and stalk people . . . it is just not safe." Amy thinks that social-network sites, in general, are getting scarier and that Facebook is in danger of becoming more like MySpace after announcing in September 2006 that it would open up registration to anyone with an e-mail address. Up until then, Facebook had restricted its user-base mainly to college students.

Amy has always been cautious in her use of social-network sites, preferring only to communicate with people she knows from school or some other familiar space. "I have a Facebook profile but not a MySpace," she said.

She explained why.

"Facebook seemed a lot safer, because when it first started out it was only college students." Referring to MySpace, Amy said, "Weird people were always trying to add me as a friend."

She told us that her future participation in social-network sites would likely be determined by how well the providers of the platforms managed access to their virtual doors. "In the years to come, I hope that there is a way to regulate who is allowed to get on . . . I would prefer to keep creepy people off."

Nineteen-year-old Loren agreed. She mentioned that Facebook once practiced a "weeding-out process," a reference to the initial decision to restrict the site to users with a ".edu" e-mail. She believes that the Facebook sign-up process discourages random friend requests and thus helps users maintain a degree of privacy.

The desire to keep undesirable people off social-network sites highlights another crucial factor in young collegians' unmistakable preference for Facebook—a strong desire for platforms that provide privacy. Contrary to popular belief, young people do care about privacy. Specifically, they care about who has access to the content and community they share with friends and acquaintances online. Even among teens who use social-network sites, a majority of them, 66 percent, say their profiles are not visible to all Internet users, according to the Pew Internet & American Life Project.

Importantly, the longing for privacy is different from the concern for security. Specifically, the Facebook users we spoke with want exclusivity rather than safety. And, like so many others, Amy and Loren's desire for privacy is also conditional, not absolute.

Amy and her peers consistently acknowledged that they like Facebook "because," as she said, "it allows people to see what you are doing." Many of them openly admitted that watching their friends' photos and keeping up with them throughout the day by way of the news feeds and status updates is what draws them to Facebook. Being watched, clearly, does not bother them. What does irritate them, however, is being watched or approached online by people they classify as outside their social sphere and below their social position. That's why many of them, including Amy, no longer use MySpace.

As our analysis of the distinctions young collegians make in their engagement with social-network sites probed deeper, our understanding became clearer. Many college students favor Facebook not only because they believe it is simple, navigable, and pleasurable, but also because it is special. In other words, the migration to Facebook is also driven by a desire to join a more exclusive community. Many of them said as much. Understood in a larger social context, it became apparent that the students we talked to preferred Facebook to MySpace for an intriguing reason: Facebook's restricted and college-oriented user-base offered an experience that was practically the same as living in a gated community; only virtual walls surrounded this exclusive neighborhood.

For eight years, Setha Low, a cultural anthropologist at the City University of New York, conducted ethnographic research on the gated community phenomenon in America. Along with examining population statistics, Low talked to home developers and families to gain better insight into the gating practices that wall off families and communities from the people and areas around them. Once associated primarily with the Sun Belt regions of the country, gated communities, according to Low, have become ubiquitous and now spread across the vast American landscape. In 1995 an estimated 4 million people lived in gated communities. By 1998 more than 16 million people called gated residences home. And the number of gated residential developments is growing. Analysis of a 1997 national survey of Community Association Institute member associations identified three types of gated communities—lifestyle, elite, and security zones. The preference for Facebook resembles the elite gates that emphasize status and privilege. Elite gates, Low explains, are "primarily about stability and a need for homogeneity."

Residents are drawn to gated communities for a number of reasons, including, as Low writes, "desire for safety, security, community, and 'niceness,' as well as wanting to live near people like themselves because of a fear of 'others.'" From a purely sociological perspective, young people's desire for privacy online is fascinating and complex. Even as they casually share some of the more intimate details of their everyday lives in a quasi-public space like Facebook, they still want privacy. In many ways, the language college students use to describe their attachment to Facebook—"safe," "private," "selective," "neat," "clean"—is amazingly similar to the language Low documents in her ethnography of gated communities. During our conversation with him, twenty-two-year-old Dylan, an avid Facebook user, candidly captured the sentiment for a gated online experience.

"When you think of Facebook," Dylan explained, "the word *safe* sort of comes to mind, because even though the criteria for membership has opened itself up to anyone with an e-mail address, you feel like it's a close-knit community of your friends." When he thinks about MySpace, Dylan said that the word *dangerous* comes to mind "because it's been open to everyone from the get-go."

When it operated largely as a college-based social site, Facebook, much more than the open-door policy of MySpace, offered its users the chance to join what Low calls a "pure space." Low explains that "the more purified the environment—the more homogenous and controlled—the greater the resident's ability to identify any deviant individuals who should not be there." And that is what Facebook does. It makes it easy to identify who should be there (college peers) and who should not (those outside the collegiate circle).

In the minds of gated residents, a pure space is also a safe place. Along with the wish to feel special, residents also prefer gates because they feel safer. Low explains that gating inevitably is about the fear of others and a world teeming with difference. But here is where gating in the physical and virtual worlds split. Facebook users in their late teens and early twenties are seldom threatened by the appearance of "others"—in this case, people not in their social sphere. Moving from MySpace to Facebook is more of a wish for prestige than protection. Being approached by someone you suspect is not in college or not middle class does not engender fright, though it does seem to intensify feelings of intolerance and indifference for people they classify as beneath their social class position.

Whereas young Facebook enthusiasts do not believe that MySpace is dangerous, there is a general suspicion that it is more promiscuous.

In fact, young men and women express dissatisfaction with the frequent spams, sex-oriented solicitations, and friend requests from random people that make MySpace, in their view, a computer-mediated community that is out of control, an unpurified space. Twenty-one-year-old Doug said "I use to have MySpace but got rid of it because it felt too open. You feel safer with a thing like Facebook . . . It doesn't feel as if you're vulnerable to the outside world. It's just your friends."

This is one of the most common complaints about MySpace—that the social mores of the site encourage random "friending" behaviors that are far less selective compared to Facebook. The nondiscriminating nature of the site is taboo among college users, a brazen violation of their expectation that an online social site restricts who comes and goes in a network. Listen to college students talk about MySpace and Facebook and the attraction to a gated experience is easily apparent.

Use of fully upgraded computers, broadband, the social Web, and other new media technologies from home or increasingly on the go are forms of cultural capital—that is, indicators of social status, tech cool, and social mobility. Likewise, the transition from MySpace to Facebook represents it own distinct bid for a social status upgrade, a step up in the digital hierarchy. I get the impression that for many college students switching from MySpace to Facebook is like moving from a modest to a more exclusive zip code. Grace, a friendly nineteen-year-old I interviewed, typifies the shift in the social-networking behaviors of young college students and, especially, their desire to belong to a more improved online community.

When I met Grace she was three-quarters of the way through her first year in college and, by that point, a clear Facebook admirer. Once upon a time, though, it was all about MySpace. "Oh, me and my friends loved MySpace when we first started using it in high school," Grace recalled. "MySpace," she said with a bright smile spreading across her round face, "was sooo cool. We used it to share pictures, see how many friends we could add, and talk about all of the crazy stuff that went on at school." And then, everything changed.

One day during the summer of 2006, after graduating from high school, Grace noticed that many of her college-bound friends, virtually overnight, had practically abandoned MySpace. Many of them had taken advantage of their new college e-mail addresses to set up profiles on Facebook. Unable to communicate routinely with her friends, Grace began to feel isolated from them and the conversations, jokes, pictures, and daily updates they shared with each other online. Even though she still saw them face-to-face, the inability to connect with them through a social-network site left her feeling like an outsider.

What Grace learned is what many students have learned—Facebook is an important source of cultural currency among the college set. In just a few short years, social-network sites have changed the bonding experiences associated with college life. During the days before the Web, the summer prior to going off to college was a period of great excitement, anticipation, and anxiety. Part of the anticipation and anxiety was not knowing things like who your college dorm mate was going to be or who was most likely to share your interests in music, movies, or a major. But all of that has changed with Facebook. Today, many students use Facebook to start making connections even before arriving on campus. Through their profiles they can identify people who share their same academic and, likely more important, lifestyle interests.

Social-network sites are not merely a source of communication among the young and the digital; they are *the* source of communication. It is the medium they use to exchange information, update each other on the latest happenings in their lives, and plan events together. Even before the first day of class, Grace was already missing out on a key collegiate bonding experience and core means of communication.

Her friends told her she had to get on Facebook because, as Grace put it, "that is the site all college students use." Once she committed to a school and mailed in her deposit to hold her spot in the incoming class, Grace requested her ".edu" e-mail account from her university's admissions office. One of the first things she did after receiving her new e-mail account was set up a Facebook profile. Speaking about the switch, Grace admitted, "It was scary for a while, feeling like I was cut off from my friends." On Facebook she and her friends exchanged thoughts about things like where to live and which fraternities and sororities to associate with. "All of my college friends prefer Facebook," she told me. "I may have one or two friends who do not use it."

I asked her if she uses MySpace anymore. "Not really," is all she said.

She and her friends had moved on. In the span of two months, My-Space was, as the saying goes, "so last year."

In MySpace, collegians see a social-network site swarming with another unattractive segment: young teenagers. College students consistently describe MySpace as immature. Nineteen-year-old Rylee expressed outrage at the presence of so many young girls on the site. She called it ridiculous. "I see ten- and eleven-year-old girls on MySpace. And they are adding me!" With more than a bit of dismay in her voice, Rylee asked, "Where are their parents?"

One of the most interesting allusions to the presence of teens in MySpace comes from one of the student researchers that I worked with. Summarizing her research, she explained that many of her peers refer to MySpace as "emo." If you are unfamiliar with the term, she described it this way. "Emo" is short for emotional and refers to the adolescent outsiders who find refuge in social sites by pouring out their heart and soul in a way that Facebook users find appallingly juvenile. Whether it is the exhilaration from a first crush or the desolation from the inevitable break up, some teens use the social Web to express their most inner feelings. Teens' use of social-network sites and blogs to grapple with the highs and lows of adolescence have often been compared to the long-standing practice of maintaining a personal journal. In the case of social-network sites, though, these personal expressions can be quite public.

Facebook's initial registration restrictions contributed to the rela-tively small community of users compared to MySpace. In reality, that made it more quaint, manageable, and most significantly, exclusive for the first wave of users. Indeed, many college students viewed the smaller Facebook as a superior platform and social-network experience next to the bulging growth and cultural anarchy they associated with MySpace. Not surprisingly, the decision by Facebook management to open up registration to anybody with an e-mail account was not very popular among college students. They understand why Facebook opened up registration—to grow as a business—but are generally unhappy about the decision.

Opening up Facebook, in their view, represented the end of an era, albeit a brief one, when the platform was largely an online hot spot for young collegians. There were no boomer parents, no "emo" teens, and definitely no future employers snooping around for incriminating pictures or wall posts. "Yeah, we [college students] still use it," twenty-one-year old Jason lamented, "but it's not the same since they opened up." A few people expressed annoyance at the rising number of applications on the site, claiming that the changes, in general, threaten to make Facebook more like MySpace—too open, too busy, and consequently too crowded.

Ten months after opening its virtual gates, Facebook experienced a rush of users. Between August 2006 and August 2007 the site experienced triple-digit traffic growth, according to Nielsen//NetRatings. The number of teens using Facebook grew even more, by 122 percent. In December 2006 Facebook had about 12 million active users. By October 2007 that number had risen to over 50 million. After registration was opened up, the Facebook user-base underwent some interesting changes. Facebook's 85 percent market share of four-year U.S. universities still made it dominant in that category. But more than half of Facebook users were outside of college, and the fastest-growing segment was among those twenty-five years and older.

Even after the initial population surge in Facebook, MySpace still hosted the biggest party in the digital universe, as more than 140 million profiles had been created since its 2004 launch. But behind the phenomenal number of profiles, a very different reality was emerging—the steady erosion of the MySpace brand among crucial parts of the young segment that is so highly coveted by digital-media entrepreneurs.

From the moment it began to blow up, MySpace's size and anarchic inclinations posed challenges for management, especially the maneuvers to monetize the site. Those challenges notwithstanding, the overall consensus was that the millions of eyeballs glued to MySpace made it a logical destination for commercial initiatives. By early 2000 there was widespread accord among speculators in the digital media economy that social-network sites were a hot property—the next new thing in the next generation Web. That is certainly what prompted Rupert Murdoch and his company, News Corporation, to acquire MySpace in 2005—the idea that the explosive growth positioned it as a new business and media model and, eventually, a cash cow in the promising but largely untapped social Web economy. But a much more bedeviling reality has emerged. A steady growing number of people use social-network sites, but as twenty-three-year old Jocelyn told us, "No one goes on MySpace or Facebook to buy things. We go there to communicate with our friends." In 2009, despite all of the hype, buzz, and cover stories, neither My-Space nor Facebook was generating noteworthy profits.

As strange as it sounds, MySpace was too successful. The platform's supersize led to a lack of quality control that began to steadily erode the user-experience and alienate key segments of young people, especially those bound for college or looking for what they perceived as a more intimate, authentic, or ad-free online community. Joel, a twenty-four-year-old music studies student, thought about the ad clutter on MySpace and echoed a common sentiment. "Have you seen those ads that pop up?" he asked. "They're basically like porn advertisements, and so I'll randomly get these and it's really annoying." The constant barrage of spams, phony profiles, ads, random friend requests, and the socially promiscuous environment undermined the value of the site for the young crowd we met, literally driving them away.

It was a rapid and stunning decline for a brand that a few short years ago was the celebrity platform in the burgeoning world of social media. In 2009, Facebook's global user base surpassed MySpace's for the first time. The demise of MySpace is a vivid reminder of just how unpredictable the fate of brands catering to the young and the digital can be.

Some of the young people we meet like Diana, a twenty-year-old Latina from the Rio Grand Valley, arrive at college from more humble circumstances and represent the select few from their schools and communities to pursue a post-secondary degree. Diana's limited exposure to new communication technologies made the transition to college much more challenging compared to that of her technology-rich peers. The soft-spoken twenty-year-old grew up in a low-income household where her family emphasized education. A regular user of social-network sites, Diana spends about four hours a day online between school, work, and play. She spoke openly about her passage from high school to a major four-year university as well as her personal history with computers and the social Web.

In school, Diana regularly scored high marks even though she never had access to her own computer. Recalling her high school days, she explained, "Every time I needed a computer, I had to go to the public library or use one at school." In both settings—a school or a public library—the ability to fashion a truly custom-fit, social Web experience is severely limited by, among other things, content and time restrictions. Most middle and high schools block access to social-network sites, whereas public libraries usually place time restrictions on users.

The complex interactions of race, class, and education have long been key predictors of household computer and Internet use. Although the technology gap between kids living in high- and low-income households has closed, a gap nevertheless persists. In a 2007 report titled *Latinos Online*, the Pew Internet & American Life Project states that 29 percent of adult Latinos have access to broadband at home compared to 43 percent of whites. Kids who grow up in broadband households are more likely to belong to a community of peers that share insights, ideas, and experiences that lead to a more robust online experience. Moreover, studies show that individuals in broadband households use the Internet for longer periods of time and for a wider range of activities than those individuals who do not have broadband at home.

During her senior year in high school, in 2006, Diana began using MySpace. Compared to many of her peers around the nation, she was a late adopter. Still, she was thrilled to join a community that by then had grown into a popular online destination for millions of people. Many of Diana's friends from school, a predominantly Latino student community, were on MySpace. They enjoyed creating their profiles for friends to see and admire. "I must have spent that whole first day creating my page and reaching out to friends who were already users."

When Diana left home to attend college, her life, as you might expect, began to change. She got her own computer—the first one she ever owned. And then, another first happened. "I discovered Facebook a week into school after I bought my laptop," she said. She immediately set up a profile and experienced an initial rush of joy that was similar to her first day on MySpace. "I had some similar excitement with Facebook, but the difference was I didn't know anyone here [at my university]," she said. "Plus, most of my friends from high school didn't attend college like I did, so I wasn't that excited in the end."

Diana realized that college students overwhelmingly populated her school's Facebook network. The only problem was that she did not come from a school that produced many college students. Most of her friends did not use Facebook. Unlike Grace and her friends, Diana had to contend with what communication scholar Nicole Ellison and her colleagues call "friendsickness," that inevitable feeling of loneliness that comes with leaving friends to attend college. While some of the earliest adopters were using Facebook to build friendships and acquaintances before arriving on campus, late adopters were missing out on this early chance to accumulate social capital and build community.

Diana's use of Facebook increased through classes and involvement in student organizations. In order to stay connected with her friends back home who were not enrolled in college, Diana continued using MySpace. Her dual use of Facebook and MySpace to navigate communities that are comprised of a cross-section of people—including the middle and working classes as well as college and non-college-bound acquaintances—is actually quite common among students.

Patty, a twenty-four-year-old college graduate and advertising professional, told us, "I use to be more MySpace-oriented, but now I enjoy Facebook a lot more. I guess because MySpace . . . I used it to keep in touch with people who are from high school and from back home, whom I don't really see much of nowadays." She said Facebook, on the other hand, is used to keep in touch with people "who are from here [where she lives and works], who I see almost everyday, or people I work with, so that's why I use it more often."

Moving back and forth from MySpace to Facebook, in instances like these, is the digital equivalent of what sociologists refer to as code-switching—a reference to the ability to master varying styles of communication and behavior that nimbly manage different cultural norms, expectations, and environments. More than anything, code-switching in the digital world illuminates the various social networks that we move in and out of. Being digital does not mean being one-dimensional. Increasingly, it means being fluid and in a constant state of evolution.

To a great extent, the social ties and personal relationships that matter most to young people are managed through digital-media technologies. Despite the differences between them, the primary purpose of platforms like MySpace and Facebook is to connect people to people—the quintessential feature in all expressions of community.

Our research suggests that virtual social ties are legitimate. Social ties and communities can even be strengthened and sustained through the use of new communication technologies. As I think more carefully about the young people we meet, two urgent questions emerge. First, what specific kinds of social capital are young people accumulating through their use of social-network sites? And second, who are they building online community and relationships with?

Robert Putnam explains that like any other form of capital, social capital can be used to achieve benevolent or malevolent ends. In his work on the collapse of community in American life, Putnam identifies two specific dimensions of social capital, what he calls "bonding" and "bridging." Writing about bonding social capital, Putnam says, "some forms of capital are inward looking and tend to reinforce exclusive identities and homogeneous groups." Bonding social capital refers to the social ties and connections to "people like us." These connections, Putnam claims, do not expand, but rather reinforce our narrower selves. And then there is bridging social capital, which, according to Putnam, is "outward looking and encompasses people across diverse social cleavages." Bridging social capital's distinct quality is connecting people who may be different in some distinguishable way—religion, class, nationality, or race and ethnicity. Networks that bridge involve connections to "people not like us" and potentially expand our horizons and our possible selves.

Does the daily use of social-network sites among the college set promote bonding or bridging? Any attempt to analyze what is happening in social-network sites must be open to nuance and change given that young people's new media behaviors are in a constant state of flux. Up to this point the evidence from our research strongly suggests that everyday engagement with Facebook tends to promote bonding and bridging. But while bridging happens, bonding is dominant.

Laura, a young college student, mentioned that most of the people in her network are middle class and white. Twenty-one-year-old Andrea said that her Facebook network "consists of the people I know closely." Andrea insisted that the longer she keeps Facebook, the more she feels inclined to keep it restricted to the people she knows very well. Also, twenty-one-year-old Matthew explained that the bulk of the people he communicates with on Facebook "are primarily my college friends, but I also interact with a lot of my friends who I grew up with and don't go to [my college]."

Among the young, white collegians profiled in this chapter, a main attraction to Facebook is the degree to which it guides online communication toward people who are likely to be young, white, middle class (in orientation), enrolled in college, and nearby. This last element is especially intriguing. Geography, we have consistently been told, no longer matters in a technology-rich environment that delivers access to anybody, anytime, anywhere. But in the initial ascent of social-network sites, the primacy of geography is undeniable. The young and the digital use social sites to communicate with people who are close by—classmates, neighbors, and acquaintances from their off-line social circles. Remarkably, as the World Wide Web grows more routine, it functions more like the local Web—a virtual network made up mainly of people in close physical proximity to its users. These and other factors embody quintessential bonding.

Twenty-year-old Nicole has been online since she was eight, making her a veteran in the world of social and digital media—online chat rooms, IMing, and social-network sites. But even though she has spent more of her twenty years of life online than off-line, it has not led to a desire to bridge—that is, to expand her social networks beyond the people she interacts with off-line. Nicole described her online network as composed of the same people she communicates with face-to-face— in her words, "upper-class Americans." She used those same types of words while characterizing Facebook as "preppy," "upper class," and "American." MySpace, she said, "is sometimes creepy, sometimes fun. It has more of a creepy element to it for me than Facebook."

She also believes that there is "probably more diversity on MySpace." This is a common view expressed by young collegians. They believe that compared to Facebook, the MySpace universe is populated with a wider variety of people. For some the diversity represented on MySpace makes the site interesting. For others, however, the demographic variety makes MySpace seriously uninviting. Results from our survey provide further evidence that bonding social capital is a common part of the Facebook experience. Young women and men alike visit social-network sites three or more times a day. Seventy-three percent of our survey participants chose "spending time with friends" as one of the things they liked most about using social-network sites. More specifically, they said they use social sites to communicate with two specific sets of friends—those that they see frequently and those that they seldom see face-to-face, presumably as a result of going away to school. This particular use of social-network sites points squarely to the preference to engage people online who are close in mind, body, and lifestyle.

Despite all of the hype about how the digital age is changing our lives, it has not changed one essential aspect of human life—who we form our strongest social ties with. Similar to life B.W.—before the Web—our most intimate bonds online tend to be formed with like-minded people. Indeed, the young people we surveyed and spoke with are attracted to online communities that connect them to people who are like them in some notable way—age, education, region, race, or class. In order to understand what is happening online, you have to understand what is happening off-line. One place to start understanding what is driving young people toward homogeneous online communities is a consideration of what author Bill Bishop has tagged the Big Sort, a reference to the geographic transformation of American neighborhoods.

The Big Sort, according to Bishop, began around 1970 as Americans underwent a massive social experiment that changed one of the most basic features of everyday life—where and with whom we live. The change in geography, Bishop maintains, is "really a sorting by lifestyle," as Americans now more than ever gravitate toward counties and communities that reflect their values, beliefs, and ways of life. Thinking further about the Big Sort and how it is remapping American neighborhoods, Bishop writes, "We have come to expect living arrangements that don't challenge our cultural expectations." Many Americans, especially the ones who enjoy a degree of economic mobility, are choosing to live near people who think, live, and look like them.

Social-network sites do not cause social divisions. The young and the digital have grown up in a world in which the geographic sorting by race, lifestyle, and ideological values—a rather extraordinary development—has become ordinary. Bishop is correct when he writes, "Kids have grown up in neighborhoods of like-mindedness, so homogenous groups are considered normal." The vast majority of the young people we meet go online to have fun by sharing their lives and communicating with their peers. And yet, the choices they make regarding who they interact with online are not immune to the social forces that are shaping their off-line lives. Like the Big Sort, the online sorting among young Facebook users is shaped by a general suspicion of difference, a split along lifestyle, and, finally, the wish to reside in communities with like-minded people.

The digital gating practices of young, white college students are especially interesting when you consider how often we hear that race does not matter to Generation Digital. In his 2008 book, *The Way We'll Be: The Zogby Report on the Transformation of the American Dream*, pollster John Zogby writes that the generation he calls the "First Globals," Americans between the ages of eighteen and twenty-nine, are "the most outward-looking . . . generation in American history." On the question that has challenged America for more than two centuries—race—Zogby says, "The nation's youth are leading the way into a more accepting future." From MTV's embrace of black pop to the selling of hip hop, this generation certainly came of age in a cultural milieu in which racial signifiers were not only visible, but also elaborately marketed entertainment. This generation personally experienced a racial milestone in 2008, the historic presidential election of Barack Obama. Liberal and conservatives alike argued that by embracing Obama's message of "CHANGE," young whites provided evidence of a generation no longer burdened by our nation's racial past. Meanwhile, our conversations with young whites show that some view MySpace and Facebook, in part, through a racially coded lens. Race, it turns out, still matters.

"MySpace," said nineteen-year-old Thomas, "is on crack. There is too much glitter and music." Twenty-two-year-old Veronica also drew a connection between MySpace and the notorious drug. "My favorite is Facebook because MySpace is absolutely on crack and overwhelming," she said. The reference to crack, a drug associated primarily with the black urban poor, is certainly not race neutral. Add to this a sentiment like "MySpace is too ghetto," and the racial marking of the digital world is apparent. Likewise, the belief that MySpace is sullied with profiles that feature "glittery, gaudy-as-shit layouts" and "too much glitter and music" invoke another racially marked term—*bling*—a popular slang derived from the larger-than-life fantasies played out in hip-hop songs, videos, and style.

Starting around the early 1990s, hip-hop music, fashion, movies, and marketing campaigns were made as much for young white consumers, especially suburban males, as black and Latino consumers. So, why would a generation that grew up consuming hip hop be turned off by the "bling aesthetics" that pervade MySpace culture? To answer that question, you have to understand the difference between "old media" (think television) and "new media" (think social-network sites).

Television and social-network sites represent two fundamentally different kinds of mediated experiences. Whereas television is about watching and consuming, social-network sites are primarily about doing and sharing. Facebook users share themselves daily through wall posts, news feeds, blogs, photos, gifts, and other activities. This kind of constant connectivity establishes varying degrees of community and intimacy. By the time they arrived in college, the late teens and young twenty-somethings we met were a little less concerned with the quantity of their online social networks and more concerned with what the quality of those networks say about them and the people they are associated with. It is one thing for young whites to listen to music inspired by the hood, and something entirely different to establish a degree of intimacy with people they believe come from the hood. This is the crucial difference between "old media" and "new media."

Whereas the use of old media platforms like television maintains distance from black and Latino youth, new media platforms like social-network sites offer a greater possibility for closeness. Back when television was dominant, young whites could consume black style and expressive culture from a distance. Social-network sites afford young whites the opportunity to interact with actual black people. However, by avoiding MySpace, the users of Facebook elect to avoid sharing their lives and experiencing a modicum of intimacy with "real" black and Latino youth in the computer-mediated spaces they frequent. Instead of venturing to bridge, some young whites choose instead to bond with each other inside their digital gates.

Social and mobile media may be changing how we connect, but as we move into the digital future, it does not appear to be significantly altering who we connect to.

In one of the first "virtual-field studies" exploring the role of race in computer-mediated social worlds, two social psychologists from Northwestern University, Paul Eastwick and Wendi Gardner, found evidence of racial bias in the online world There.com. The experimenters created two avatars, one with light skin and the other with dark skin. Modeling one of their investigations on a classic "compliance technique" experiment called "Face-in-the-Door" condition, Eastwick and Gardner made a large request of There.com participants that was sure to be refused. It was the response to the second, more moderate request that the researchers were really interested in. Previous studies found that the moderate request usually leads to greater compliance from study participants. It turns out that the participants believe that the requester has made a concession and, thus, are more likely to reciprocate. Researchers believe that the participant's decision to agree is based on their assessment of the requester.

As expected, Eastwick and Gardner noticed that the more moderate request led to greater compliance. But when they examined the responses to the light- and dark-skinned avatars, a statistically significant difference emerged. Among the light-skinned avatars, 20 percent more people said yes to the second request. Among the dark-skinned avatars, only 8 percent more of There.com participants said yes. The researchers concluded that the social influence of race might have been a factor, though they were unsure of the precise nature of the effect. Did participants respond based on their perception of the avatars' appearance (skin color) or their perceptions of the person controlling the avatar? Either way, Eastwick and Gardner write, "the virtual world may not prove to be a perfect utopian getaway from the real world." The racial perceptions and biases we develop in our off-line lives, they conclude, likely creep in to our online lives.

More than a decade after that famous *New Yorker* cartoon, there is an emerging body of research that suggests that as we travel deeper and deeper into the online world, being a dog in the off-line world matters after all.

FIVE

We Play

The Allure of Social Games, Synthetic Worlds, and Second Lives

Video games are a great social connector, and allow for making friends.

—Chase, twenty-one-year-old college student

When Nielsen Media Research released its prime-time viewing data for the new fall season in 2003, the major television networks were reminded, once again, of their most vexing challenge: delivering young male viewers to their advertisers. That fall, Nielsen reported that prime-time viewing among young men was dropping precipitously. Compared to the year before, viewing among men between the ages of eighteen and thirty-four fell 12 percent. Even more alarming was the decline among younger men between the ages of eighteen and twenty-four—20 percent fewer of them were watching television during the prime-time hours. The television industry found itself caught in the middle of a real-life mystery as Nielsen's data left it scrambling to figure out what the press began calling the "missing young men story."

Throughout TV land, the idea that young men, a lucrative and heavily sought-after demographic, were abandoning television in droves caused a major stir. A *New York Times* article likened Nielsen's viewer ratings to "a nuclear strike, a smallpox outbreak and a bad hair day all rolled into one." Total disbelief was the primary response from television industry executives. "Frankly what we're seeing strains credulity," said Alan Wurtzel, the president of research at NBC. David F. Poltrack, executive vice president for research at CBS, explained that while the claim that young men's television viewing was on the decline was defensible, the trend, in his view, "should have been seen gradually over time," and not "all of a sudden." Wurtzel was even more forceful in his criticism of Nielsen. "You never see these kinds of unanticipated declines," Wurtzel said. "It's like saying that all of a sudden all these guys have completely turned off the use of TV. It does a huge disservice to the entire TV industry." A Nielsen twelve-year trend line showing a consistently downward slope in the television viewing behaviors of young adults suggested that the decline was not nearly as sudden as it appeared.

In reality, the television industry was coming face-to-face with the aftershocks of the rising number of broadband homes and young people's move to digital. Still, the industry men's disbelief was understandable. Ordinarily, behavior as habitual as television viewing did not change so profoundly so quickly. But the digital age is no ordinary time. By 2003 television viewing as the industry had known it for more than fifty years was a relic, a thing of the past. Young people still watch television but, as pop culture critic Douglas Rushkoff notes, in ways that are strikingly different than previous generations. They watched online or while they were doing other things, like Facebooking, downloading music, uploading pictures, or sending instant messages.

The "missing young men story" was interesting, but Nielsen's television viewer data only told part of the tale—more intriguing is the question, what are young men abandoning television for?

Peter Daboll, president of comScore Media Metrix, an Internet audience measurement service, stated that, "The fact that more than 75 percent of 18–34 year-old men in the U.S. are using the Internet seems to take at least some of the mystery out of the decline in TV viewing among this prized demographic." In short, all you had to do to find young men was go online. Some of comScore's key findings on Internet use that September (2003), the month that the networks began rolling out their fall line-up of new shows, is revealing. The roughly 27 million eighteen- to thirty-four-year-old males who used the Internet in September averaged about thirty-two hours per person online. That was 17 percent more than the twenty-seven hours the average Internet user spent online during the month. After reporting that young men viewed, on average, in excess of seven hundred more pages than the average Internet user, comScore called them voracious consumers of online content.

Young men have consistently been among the early adopters, explorers, and users of new communication technologies. No demographic has taken greater advantage of the on-demand capabilities of the new media environment than young males, the most wired segment in America. In our survey, young men downloaded music significantly more than young women. They also consumed more online video content than any other group. In a 2007 report titled *Online Video*, the Pew Internet & American Life Project found that young men were more likely than any other group to report a preference for both professional and amateur online video content. But one technology rises above all others when it comes to grabbing young men's attention—games.

Nielsen explained that young men, ages eighteen to thirty-four, were spending 33 percent more time playing games compared to a year earlier. In homes across America, interactive entertainment is on the rise. The growth of the games industry is verification of that. Between 2004 and 2006 there was an 18.5 percent expansion of U.S. households—38.6 million to 45.7 million—with televisions that also have video-game consoles. The sales of games remain solid even as other entertainment industries—music, television, and film—struggle to reinvent themselves in the digital age. According to the Entertainment Software Association, "U.S. computer and video game software sales grew six percent in 2007 to $9.5 billion—more than tripling industry software sales since 1996."

A study by Nielsen Entertainment found that active gamers—persons who spend at least one hour a week playing games—spend a considerable portion of their money and time on gaming-related activities. Active gamers spend about $58 a month on entertainment. Roughly 28 percent, or $16, of that is devoted to games. Equally significant is the amount of their leisure time devoted to games. About a quarter of active gamers' weekly leisure time is spent on games, or 13 out of 55.3 hours. Michael Dowling, general manager at Nielsen Interactive Entertainment, believes that "as games continue to increase its share of entertainment leisure time, it's quite possible playing video games will assume a significant role as a common cultural experience, in the way that movies and television do today." This is no longer a possibility; it is a reality.

A growing number of young men are turning to interactive entertainment like games rather than television and movies as their first source for leisure and a desired choice for social interaction with their friends. Still, to fully comprehend why games are stealing away young male eyeballs from TV land, you have to look beyond game stats or the revenue the industry generates to the kinds of experiences it facilitates.

In our discussions with young men, the allure of games is indisputable. We asked users of massively multiplayer online role-playing games (MMORPGs), "Which do you prefer, an evening playing games or an evening watching television?" The answers we received from a group favorably predisposed to games were not surprising. Predictably, nearly everyone we spoke to chose games. Still, their reasons for doing so offer insight into the changing media behaviors and attitudes of young men. Whereas a few of them acknowledge that sports programming and DVDs can be entertaining, most indicated that television was simply too boring and in their view too passive.

"Games and television," twenty-two-year-old Nelson told us, "are totally different. With television, you sit there, you watch. With games you are playing. You are doing things." Another gaming enthusiast, Lee, expressed a similar sentiment. "I feel like I'm wasting more time if I'm watching TV with all of the commercials and such." Lee added, "With games I feel like I'm accomplishing a goal. With Ventrillo [a voice technology that allows real-time chat while game playing] I feel like I am just hanging out and talking with friends." Comments like these underscore the immense challenge television faces, as well as why games are so perfectly suited for a generation of young people who prefer doing rather than watching when it comes to their media use.

Sit and talk with young people today about media and communication technology and a fascinating generational ethos comes into clear view: the idea that they are not simply consumers of media but also creators and participants in media. Peter, a twenty-one-year-old college student, described games this way: "It's more than just sitting in front of a screen just having stuff thrown at you. You are interacting in a story.

I find that there's more in a game than in television." John, a nineteen-year-old student speaking about the allure of games, said: "Instead of sitting back and watching a plot happen, you actually get to participate in the action."

In the rich and textured three-dimensional worlds common today you can go anywhere and be anybody. Ultimately, no matter if it is the promise of virtual wealth, the temptations of self re-creation, or some other innovation, games embody what Katie Salen and Eric Zimmerman call "culturally transformative play," a feature that renders television old, inactive, and increasingly irrelevant.

Equally troubling for the television industry is the fact that games offer a totally immersive experience. Intense game play demands a kind of focus and attention that is increasingly rare in today's new media environment. Unlike other social media activities such as social-network sites or IMing, games are much more resistant to the multitasking ways so common in today's media environment. When most people play games they are rarely, if ever, downloading music, streaming online video, Facebooking, or watching television. The fact, too, that some of the more popular games are incredibly social also explains their rise above television as a preferred leisure activity among the young and the digital.

In its third annual Active Gamer Benchmark Study, released in 2006, Nielsen Entertainment announced that the social elements of video games are becoming an increasingly important part of the overall gaming experience. Active Gamers, the study reports, spend upwards of five hours a week playing games socially. Teenagers, according to the study, are socially involved in gaming about seven hours a week. What is social gaming? Social gaming occurs in various styles and subgenres of games but refers primarily to those games in which social interaction with others is a key component of the gaming experience. Many MMORPGs are designed to be social, insofar as they encourage interaction between players, often in the form of collaboration and teamwork. Users of MMORPGs are drawn to the platforms precisely because they afford the opportunity to engage other users in real time.

Among the four-pack, another aspect of social gaming is more evident—getting together with friends, face-to-face, to play games together. For young men like Derrick, Chase, Trevor, and Brad, games are more than a leisure activity; they are, quite simply, a social activity. This particular use of games is emblematic of the constantly evolving role that social and interactive media occupy in the lifestyles of young people.

Chase's media journal is quite candid about how his use of games has changed in recent years, going from a primarily single-player experience to a multiplayer experience. "When I initially started," Chase wrote, "I played games because they provided me a way to fill time." Throughout his school-age years, Chase played games often, but he generally played them alone. During his final year of high school, he logged in about twenty to thirty hours a week playing games. Though he played soccer and hung out with his friends, he said "there would be a lot of time during the week that I would have multiple hours with nothing to do, so I just played video games." It was around this time that Chase began playing *World of Warcraft* (*WoW*).

In MMORPGs, completing a single mission could take as much as one to two hours. Addressing the open-ended structure of MMROPGs, Chase recalled, "There is always something else to do and achieve so you never come to an actual end to the game." Playing these games meant that Chase was also spending greater amounts of time in front of his computer. He enjoyed meeting other players online, but in retrospect, believes the use of massive online games reduced the time he spent with his physical-world friends.

One of the things that we noticed in our work with nineteen- to twenty-six-year-olds is that their media behaviors change in some striking ways, as was the case when Chase began college. He watched less television, and while games were still his primary outlet for leisure and entertainment, he did not have nearly as much idle time on his hands. Moreover, the style of games he played changed, as did his reasons for playing. During the five months that I spent getting to know Chase, he was playing games like *Halo 3* and *Guitar Hero III*. Rather than play games to fill time, a largely individual endeavor, Chase said he plays to spend time with his friends, a largely social endeavor. In his journal,

Chase wrote, "These games [*Halo 3* and *Guitar Hero III*] provide a community aspect and a way to connect to people."

Chase's desire for social gaming experiences is, of course, consistent with the transformations that are reshaping the culture of gaming. During our next one-on-one conversation, I made sure to follow up on his revealing journal entries. Chase told me that he is beginning to appreciate games for a whole different set of reasons. "Video games are a great social connector, and allow for making friends," he told me while we were sitting and talking one day in an outdoor courtyard. Chase and the other members of the four-pack began their journals right around the time the much-anticipated *Halo 3* was released. To ensure they would receive copies of the game, they stood in line at midnight on the evening *Halo 3* went on sale. When they arrived home at about two o'clock that morning, they played the game until sunrise.

Over the next four days Chase and his friends played *Halo 3* almost nonstop. "We played against people that we knew and then against other people online who we did not know," Chase remembered. *Halo 3*, he explained, also became a bridge for meeting other people in his dorm. "You could be walking down the hall and pass a room full of guys playing *Halo* and before you knew it you had been invited in to join a group that may have included guys you never met before." During the first few days of playing *Halo 3*, Chase bounced from one group to the next. In addition to nourishing previously established friendships, he planted the seeds for a few new acquaintances.

Curious to know if it was the game or the social interactions it facilitates, I asked Chase, "What is the main appeal of a game like *Halo 3*?"

"Part of it is the game itself," Chase acknowledged.

Many of the guys had played and knew the two previous versions well. With *Halo 3* came the pleasure in Chase's words "of re-learning a game that you already knew." There was the excitement of mastering new maps and discovering where weapons and other game-related items were located.

"But another part," Chase added, "is the fact that you meet people through it." The three-on-three and four-on-four matches make for great fun and social interaction, online and off-line. During the five

months I was with the four-pack, the games that they mentioned frequently such as *Halo 3* and *Guitar Hero III* had one thing in common: a
strong social dimension. Nine of the ten top-selling games in 2007 were
games that had multiplayer capabilities. In many of these games, the fun
is as much in the social interaction and camaraderie as it is in the game
itself.

Another big hit in the dorm the four-pack called home was Nintendo's Wii gaming system. Much more than its chief competitors, Xbox
360 or PlayStation 3, the Wii is synonymous with social gaming.

Shortly before resigning from his position as president of Sony's world-
wide game studios, Phil Harrison discussed the impact of social gaming
in his industry. At a private lunch hosted by game developer David Perry
during the 2008 Gamed Developer Conference in San Francisco, Harrison acknowledged that Sony's development of PlayStation 3 was a serious strategic mistake: "It's a very interesting and frustrating thing for
me to experience because I have been banging the drum about social
gaming for a long time," Harrison said. "And our Japanese colleagues
said that there is no such thing as social gaming in Japan: 'People do not
play games on the same sofa together in each other's homes. It will never
happen.' And then out comes the Wii."

In the run up to the 2006–07 next-generation console war, Sony
and Microsoft bet on big graphics and big power with PlayStation 3
and Xbox 360, respectively. Sony billed the PlayStation 3 as the world's
most sophisticated gaming system. And virtually everyone agrees. The
console has many of the leading technical benchmarks—high-definition
graphics, unsurpassed processing power, and a Blu-ray Disc player—
that made it the most powerful gaming system ever made. Early on,
though, analysts wondered if the console was over-engineered in such
a way that only hard-core gamers could appreciate. One review wrote
that the system "feels like a brawny but somewhat recalcitrant special-
ized computer." And in Japan, Sony's home turf, an analyst for Nomura
Securities explained that the PlayStation 3, "for many consumers, it's
still just too much, too much machine for too much money."

With PlayStation 3, Sony produced a console that delivered an intense but not necessarily social gaming experience. By contrast, Nintendo's Wii console delivered a more modest machine—in terms of price, power, and graphics—but one that also appeals to what the industry characterizes as casual and social gamers.

Many of the earliest reviews of the Wii raved that the console encourages people to play along with friends, family, or in groups, thus creating an atmosphere that is as much about gathering as it is gaming. Like other kinds of home-based media and entertainment devices, most notably the television, video games have generally been viewed as a solitary form of entertainment. The image of a socially isolated, sedentary, blurry-eyed teenage male furiously working a joystick has been the dominant caricature of gamers. It is easy to forget that early on in the gaming console's rise as a source for home entertainment that families actually played together. Throughout the 1990s, however, games grew more challenging, youth obsessed, and male centered. In his perceptive review of the console, Seth Schiesel, a *New York Times* games reporter, writes that "the Wii is meant to broaden the gaming audience yet again and turn on that broad swath of the population that got turned off by video games over the last two decades."

Analysts and even Nintendo officials proclaim that the Wii's greatest influence is making gaming fun and accessible for a wide cross-section of people—young and old, male and female, casual and serious. When my seven-year-old daughter received a Wii as a gift, her sixty-plus-year-old grandfather, a man who had not picked up a game control in nearly twenty years, was drawn to the action. The console and the innovative remote make gaming fun, inclusive, and intuitive rather than serious, exclusive, and intimidating.

The selling of the Wii is also instructive. According to Schiesel, "In an entirely counterintuitive, brilliant move, most of Nintendo's ads are now shot from the perspective of the television back out at the audience, showing families and groups of friends having fun together." Schiesel maintains that Nintendo "realized that emphasizing the communal experience of sharing interactive entertainment can be more captivating than the image of some monster, gangster or footballer on the screen." Nintendo's move was crafty and timely. Social gaming, you see, is illustrative of a broader cultural transformation: the rise of social media experiences as the hallmark feature of how we use communication technologies in our everyday lives.

In their own unique way, each member of the four-pack talked a lot about games as both a social lubricant and a social glue. The former refers to how games can make it easier to strike up conversations with new acquaintances, while the latter is a reference to how games give established friends a fun way to grow closer to each other. It is, in the end, the social aspects of gaming, more than anything else, that make games the most preferred source of leisure for the four-pack. Games are an integral part of their social life and social networks. Throughout the week they are constantly getting together with friends and acquaintances to play games. I asked the four-pack how close they were to their off-line circle of gaming partners.

"Oh, some of these guys," Derrick explained, "will be my friends for life." Trevor explained how their affinity for each other reaches beyond games. "We go out and get food a lot. We go to concerts together and play intramural sports too."

Notably, the four-pack are not slaves to their consoles and PCs. Still, games for the young men I got to know are the basis for forming a social network made up of both relatively strong and weak ties. As Chase told me during one of our conversations, "The relationships are being formed by the games, but strengthened by all of the different activities we do together."

Social gaming for young men like those of the four-pack primarily involves getting together with friends and acquaintances, face-to-face, and playing games. But for millions of others around the world, social gaming comes in a very different form: logging into a massive 3-D computer-generated world to play, interact, and occasionally bond with people they, in all likelihood, will never meet face-to-face.

With an estimated 10 million subscribers (at the time this book went to press) *World of Warcraft* (*WoW*) is planet Earth's most populous MMORPG. Launched in 2004, *WoW* ascended rapidly as a destination of choice for gamers drawn to the epic scope and anything-is-possible atmosphere common in what games scholar Edward Castronova calls synthetic worlds—"crafted places," he writes, "inside computers that are designed to accommodate large numbers of people." We spoke with several users of *WoW*. In our conversations we focused less on game-play machinations and metrics—how much gear they owned, what they do to enhance the reputation or level of their avatar, or the details of a raid they joined. Our goal instead was to learn more about gaming as a particular dimension of the social- and mobile-media lifestyle. Gaming in massively multiplayer online role-playing worlds has interesting consequences for life outside the virtual world.

Among the synthetic-world users that we spoke with, none was more forthcoming than twenty-one-year-old Curtis. Over the course of a four-month period, we learned a lot about Curtis and how his migration to digital is most evident in his insatiable appetite for games, especially *WoW*. Games are Curtis's most prominent source of leisure and a lifeline to others and the world around him. Most of the *WoW* users we spoke with are extremely active, spending, on average, between fourteen and twenty-eight hours a week playing the game. In a 2005 large-scale survey of MMORPG users, it was reported that the average user spends nearly twenty-three hours a week in their chosen game. That same survey found that 61 percent of "respondents had spent at least 10 hours continuously in an MMORPG." Eight percent of users spend what amounts to a forty-hour workweek in massive games.

I asked Curtis how many hours a day he plays *WoW*.

"Oh my, too many," he said sheepishly, before answering, "I'd say on an average day, eight to ten hours."

Around seven o'clock each evening, Curtis powers up his computer, grabs an energy drink, and logs into *WoW*. By the time Curtis usually logs out, it is close to two or three o'clock in the morning. His first experience with a massively multiplayer online role-playing game was *Counterstrike*, one of the first successful online multiplayer games. "I started playing that when I was sixteen," he said. His dad worked on computers all of the time and his mom played games. "She was an absolute wizard at Tetris," Curtis said proudly. He and his mom constantly played games together. When we met Curtis, he had been playing *WoW* for two years.

Like most *WoW* enthusiasts, Curtis thoroughly enjoys the social aspects of the game. "I have a lot of friends in *WoW* and it's fun to interact with them," he told me during our initial conversation. Indeed, the best virtual worlds are lively and social—places where people from all over the world visit and interact with others on a regular basis. Curtis described himself as a bit shy but believes *WoW* "promotes a lot of social interaction."

Another aspect of Curtis's fascination with *WoW* is attributable to the fact that the virtual world is characteristic of a recurring but relatively recent development in game design—open-ended or free-form gaming. In synthetic worlds like *WoW* the design elements and rules of play are extremely open and flexible—qualities that enable participants to create a more distinct and personalized experience. If you do not feel like slaying a monster today and would rather chat with some of your gaming comrades—no problem. Want to be super competitive and increase your avatar's level, power, and reputation? You can do that too. *WoW*, as Nicolas Ducheneaut and his research associates argue, is "two games in one." While some players pursue a leisurely, individual, and single-player experience, others pursue a more intense, social, and multiplayer experience.

Some game scholars compare "free-form" or open-ended gaming to playing in a sandbox. What does this mean? Visualize a kid playing in a sandbox. Because she is not bound by any established set of rules, her playtime is free, open, and enormously creative. Conditions like these permit her to make wonderful creations, discoveries, and, yes, even mistakes. It is this particular aspect of open-ended games— exploration—that leads some scholars to characterize them as excellent learning environments. It is the idea that you learn in games not by being told what to do but rather by doing. Trial and error encourages discovery. Learning in games is a hands-on experience. Kurt Squire, a games scholar, refers to open-ended games as "possibility spaces"—that is, spaces that invite users to imagine, create, and build new identities and worlds. Curtis underscored this aspect of *WoW* when he said, "It is fun in that you can do things that you would normally never be able to do."

Logging in to *WoW*, Curtis believes, is analogous to reading a fascinating book. "The storyline is so gripping, it can carry you away," he said. Like a number of fantasy-based games, the story in *WoW* is constantly evolving. According to game scholars Edward Schneider, Annie Lang, Mija Shin, and Samuel Bradley, story in video games matters. "Story," the authors write, "is something that video game players enjoy; it helps involve them in the game play, makes them feel more immersed in the virtual environment, and keeps them aroused." In addition to the lore and published novels manufactured by Blizzard are the elaborate narratives that *WoW* users write and post online.

WoW is the quintessential participatory platform, a model of the interactive media franchise that is poised to thrive among a generation of young men and women who prefer using Web-based tools and applications to create their own content and build their own world. It represents what scholar Henry Jenkins calls "convergence culture." Social-media platforms—MMORPGs, wikis, blogs, and social-network sites—have triumphed over television because they offer participants the opportunity to tell their own stories. In the days when old media regimes like print, movies, and television were dominant, the media content we consumed was typically limited to the stories and content created by professionals. In the new media regime, users expect to be able to create and control their own content and increasingly create their own stories.

Despite the "cult of the amateur" criticisms by some, today's young media users will have it no other way. As a participant in a good role-playing game, you are not simply watching the story unfold; you help determine the outcome of the story. You are not simply pulling for the hero; you are the hero. You are, all at once, the writer, director, actor, and audience. It is an absolute reinvention of what it means to be a media consumer and one of the truly revolutionary aspects of the migration to digital.

Another fascinating aspect of life in the virtual world is the creation of an avatar, a digital self-representation that users maintain and present to others. Think of an avatar as the character, the virtual persona and embodiment of a desired or re-imagined self. Avatar-mediated play is the medium through which users not only fashion a second self but a second life too. As one twenty-one-year-old synthetic world user says about MMORPGs and the maintenance of her avatar, "I like that I can be somebody else for a couple of hours."

Curtis absolutely enjoys the identity creation aspects of *WoW.* During our conversations with Curtis he was managing three avatars in *WoW* but has played as many as nine characters with varying degrees of interest. At the time of our conversations he was especially enamored with Anna. His inspiration for the avatar came from a popular science-fiction television series based on a crew that survived losing a civil war and then found themselves as pioneers in a new galaxy. Curtis had even created an elaborate backstory for Anna. "She has now become an elf who was raised in human lands," he explained. There are a number of synthetic-world users who travel to great creative lengths to construct and enliven their avatars. In addition to designing the look of their avatar, synthetic-world users also create elaborate backgrounds, identities, and role-playing repertoires for their digital self-representations.

Curtis's decision to create a female virtual self is actually quite common in fantasy-based computer-generated worlds. Researchers call this gender swapping, or to use the words of MIT professor Sherry Turkle, "virtually cross-dressing." There is a long history of playing around with identity, especially gender, in online worlds. Men present as women, and, similarly, women present as men. Men and women gender swap for complex and often distinct social and psychological reasons. For some women, presenting themselves as men in a synthetic world can license them to exercise more power and authority. Frustrated with never being taken seriously as a woman in *EverQuest*, Becky, a petite Asian American woman, swapped gender and race. "I became the biggest black guy I could find," she said. "When I play this big guy, everybody listens to me." The male avatar was a new and invigorating source of power. "Nobody argues with me. If there's a group of people standing around, I say, 'Okay, everybody follow me!' And they do. No questions asked."

Men present themselves as women for a variety of reasons too. In her discussion of men who practiced gender swapping in some of the first computer-generated virtual worlds, multiuser dungeons (MUDs), Turkle explains that for some men, becoming a woman in a virtual world was a powerful and evocative experience. Some of the men Turkle

met presented themselves as women because MUDs afforded an opportunity to break away from the rigid roles and cultural expectations of
masculinity. One man she profiled, twenty-eight-year-old Garrett, explained his decision to role-play as a woman this way: "I wanted to know
more about women's experiences . . . I wanted to see what the difference
felt like. I wanted to experiment with the other side." Dissatisfied with
conventional gender norms, Garrett said, "As a man I was brought up to
be territorial and competitive. I wanted to try something new."

For Turkle, gender swapping holds the promise of social transformation. She compares virtual cross-dressing to what anthropologists
call *dépaysement*—which means to immerse oneself in a different culture,
indeed, a different world. Upon returning to your culture, you see the
world through a different lens. So, by identifying with women in virtual
environments, men might begin to see the world and the challenges
they face from a different perspective. Still, others dispute this observa-
tion, claiming instead that gender swapping is rarely if ever transformational. In most cases, critics point out, cross-dressing in the synthetic
world reinforces gender stereotyping and thus redraws rather than
erases gender boundaries. This is especially true among men who create
supersexual creatures and body types—with large breasts and impossi-
bly slim waistlines—that many heterosexual men find attractive, but also
submissive. Rather than engage alternative notions of femininity in the
virtual world, men tend to embrace the established gender norms and
conventions that reduce women to a marginal status and sex objects.

When I asked him about his decision to gender swap, Curtis explained it this way. "The males [in *WoW*] look like complete idiots
. . . the females look a lot cooler."

Curtis described Anna this way: "She is very nice, very upbeat, very
cheerful, very helpful, and more than willing to lend a hand." He went
on to say, "She is also very respectful of those who are of higher rank
or more experience than her." Anna, according to Curtis, was a soldier
practically all of her life, so he tends to play her as very respectful, helpful, and dedicated to the welfare of everybody.

During one of our conversations I asked Curtis if there is a difference between him and the avatar he maintains with such great care and
imagination.

"Yes, there is a definite difference," Curtis said. Then, he immediately added, "Actually, a lot of me does tend to come out in the character, I personally see myself as very upbeat and respectful, very kind and willing to help other people out."

"Curtis," I inquired, "do you use your avatar to express things or parts of yourself that you may not feel comfortable expressing in the real world?"

"I don't doubt that I use her to express some of my more feminine qualities that I'm typically not comfortable with or are a little less acceptable to express in the real world."

"Can you give me a specific example?" I asked.

"Caring about how she looks, dressing her up, and having little nice outfits, which aren't exactly strong masculine qualities."

Curtis also confessed that Anna is more confident than he is. She is, in his words, "someone who knows what she wants and is very determined to achieve it."

Synthetic-world users like Curtis take their avatars seriously. Outside observers might say too seriously. But maintaining a virtual persona is more common than we realize. The truth is, more and more of us manage a virtual or computer-mediated persona of some sort. In some cases it may be in the form of a personal profile on a social-network site—think Facebook or LinkedIn. The pictures and videos we upload, the friends we accept, and the personal data and commentary we post tell interesting stories about either who we are or who we want to be. Also, some individuals maintain a virtual persona in chat rooms—anonymous online spaces that allow us to disclose personal details with few, if any, risks. Bloggers fashion a distinct digital persona based on the content they produce for their readers. And then there are the avatar-mediated forms of identity creation that take place in the virtual world.

The rules for making digital identities vary significantly across different platforms. Maintaining a digital self in Facebook, for instance, is very different than maintaining a digital self in *WoW*. Whereas users of Facebook engage in identity management and self-representation, users of synthetic worlds engage in identity play and self-experimentation. In most Facebook networks there is an expectation that the person you present closely approximates the person you are in the physical world. Identity making in that platform is about managing and presenting a self that off-line friends and acquaintances know and see. By contrast, there is little to no expectation that the person you present in a role-playing game will resemble the person you are in the physical world. For that reason, building an avatar for the virtual world is much more explicitly playful and experimental—it is a license to build what some refer to as an alternative, imagined, or second self.

As the number of computer-mediated personas rise, the research on the social lives of avatars grows more fascinating. In their examination of life in the virtual world, researchers Nicholas Yee and Jeremy Bailenson set out to test the validity of a theory Yee calls the "Proteus Effect." Essentially, the theory maintains that the avatar—how it looks and how it is perceived by the user—will influence the user's behavior. Proteus, a sea god in Greek mythology, is known for his ability to change form. The word Proteus is the root for the adjective *protean*, which means versatile, mutable, and flexible. These, of course, are traits closely associated with identity in the digital age.

Yee and Bailenson conducted two experiments. In both experiments the participants were shown a mirror image of their avatar for roughly sixty to seventy-five seconds, just before they interacted with other avatars in a Collaborative Virtual Environment. In the first study, some study participants were assigned attractive avatars. The other set of participants received unattractive avatars. Participants were then asked to interact with an avatar of the opposite sex in a virtual environment. What did Yee and Bailenson learn from this experiment? Among other things, they noticed that participants who were assigned more attractive avatars moved significantly closer to the opposite sex. Also, when the confederate avatars asked the study participants to "tell me a little bit about yourself," followed by "tell me a little more," Yee and Bailenson noticed something interesting. The attractive avatars were much more willing to disclose information about themselves than the participants who were assigned less attractive avatars.

Assessing the results, Yee and Bailenson conclude, "The attractiveness of their avatars impacted how intimate participants were willing to be with a stranger." Most important, the attractive avatars exuded more self-confidence than their less attractive counterparts.

Intrigued by the results of the first experiment, the Stanford researchers conducted a second study. One group of participants received tall avatars while another group received shorter avatars. The behavior they measured in this experiment was based on a negotiation task called "the ultimate game." In this game, two individuals participate by taking turns deciding how a pool of money will be split between them both. After one person suggests a split, the other person can accept or reject the offer. If the participant accepts the split, the money is shared accordingly. In the case of a rejection, neither participant gets a share of the money.

Yee and Bailenson asked the participants to split $100. Once again, they found support for the "Proteus Effect." Participants assigned taller avatars were more willing to make unfair splits than those who had shorter avatars. Participants with taller avatars were also more likely to reject unfair offers than their shorter counterparts. Like the previous experiment, participants assigned taller avatars displayed more confidence and greater self-esteem in the game than those assigned shorter avatars. "These two studies," Yee and Bailenson write, "show the dramatic effect that avatars have on behavior in the digital world." According to Yee and Bailenson, the self-representations chosen by synthetic-world users has a decisive influence on how they behave in the virtual world. Their behavior mutates in relation to how their avatars look and consequently how they perceive themselves in the virtual world.

Most people intuitively grasp how an individual's qualities in the physical world might influence their decisions and behaviors in the virtual world. A short and bashful male, curious about experiencing the world as a strong masculine creature, self-presents as tall and powerful in a virtual world. A person who feels unattractive in the off-line world creates a dazzling-looking avatar. A recluse, eager to live life as a dynamic socialite, becomes a hip club owner in the 3-D virtual world Second Life. You get the idea. But contemplating the opposite—how our decisions in the virtual world shape our behavior in the physical world—is much more difficult to grasp. Is it possible that the person we create in the virtual world can influence the person we are in the physical world? Given the widespread adoption of virtual personas, it is an issue that, in all likelihood, will become increasingly debated as we grapple with the consequences of "being digital" as a way of everyday living.

Currently, some researchers point to the formation of guilds, those player associations that form in massive online games, as a sign of the virtual world influencing the physical world. Guilds, as we learned in our interviews, can be made up of several members committed to executing extremely challenging tasks. Managing these kinds of player associations—both the mission and the members—requires delicate coordination and detailed communication. Skills, some argue, that are transferable beyond virtual-world borders. Yee reports that "10% of [MMORPG] users felt they had learned a lot about mediating group conflicts, motivating team members, persuading others, and becoming a better leader in general." Yee goes on to note that 40 percent of users believed they had learned a little about these skills. Elsewhere, John Seely Brown and Douglas Thomas contend that "the process of becoming an effective *World of Warcraft* guild master amounts to a total-immersion course in leadership." It remains to be seen if humans will deliberately use the virtual world as a place to test skills, talents, and behaviors intended for the physical world.

Becoming someone else in the online world can force virtual-world users to confront the person they are in the off-line world. But even then there are no guarantees that the physical-world person will be transformed in some dramatic way.

"Sometimes," Curtis told me, "I find myself doing things with Anna because I personally want to do it, or I'll sometimes find that I'm thinking as Anna [in the online world] would think, so I really get into it."

Earlier in our conversation I recalled how Curtis mentioned that Anna is much more confident than he is. "Does that make you more confident?" I asked.

"No, I don't think so," he replied. "Anna is a more confident and courageous person than I am."

Thinking out loud, Curtis added, "Maybe one day, I'll be as confident as Anna, but that day has not arrived yet. In the real world I know that I am not Anna; I know that she can do things that I can't."

■ ■ ■

Along with producing and managing online identities, users of synthetic worlds also find themselves participating in online communities. The best virtual worlds are vital worlds. Places, in other words, that provide constant opportunity to meet, interact, and build relationships with others. Not all online places, however, offer the same kinds of opportunities for community. Take, for example, the kinds of community one is likely to find in Facebook versus *World of Warcraft*.

A frequent user of Facebook logs on several times a day to stay connected to people she knows or sees regularly. By contrast, a user of *WoW* tends to draw excitement from interacting with people he may only know in the online universe. The Facebook user goes online not to escape her off-line life but rather to engage and enrich her off-line life. Every so often the *WoW* user visits the virtual planet in order to gain momentary relief from his off-line life, seeking, instead, to enrich his online life.

Talk to synthetic-world users and most will tell you that they are drawn to the virtual because of the chance to meet and interact with other people. In the past, critics of computer-mediated communication generally dismissed the idea that it was possible to bond with strangers in a virtual environment. Even today, many find it difficult to believe that a meaningful relationship can form online between two people who have never met face-to-face. But humans have been connecting and bonding in virtual environments for almost three decades. The real issue is not that massive online games are not social. Many of them absolutely are. While you can explore planet Azeroth, the fictional universal setting in *WoW*, in a single-player mode, the users who stick with the game, eventually, tend to favor the multiplayer mode that encourages social interaction and collaborative forms of play. Unlike single-player games, you advance through the upper levels of *WoW* not by individual mastery of the game but rather by cooperating effectively with others, often through player associations called guilds.

The more interesting question is not whether or not the virtual bonds and relationships formed in places like *WoW* are real, but rather, what is their value to the millions of people that build, experience, and maintain them?

Erase the computer-mediated aspects of virtual bonds and you could make a credible case that they parallel the traditional strong and weak tie relationships that I discuss in chapter 3. Similar to face-to-face relationships, virtual bonds are extremely complex social networks that come in varying degrees of intensity. Some may find it difficult to classify the social interactions in a MMORPG as classic "strong ties" because, admittedly, building intimate connections in a fantasy-based, role-playing environment is challenging. It is not necessarily a question of whether or not synthetic-world users are bonding but rather with whom they are bonding—each other, their avatars, or something in between. Still, some of the connections established in the virtual world certainly qualify as more than weak ties, which often imply infrequent or dispassionate interaction. Yes, the interactions in virtual worlds are computer mediated, but they can also be frequent, extraordinarily detailed, and at times emotionally intense.

Virtual-world users speak fondly and frequently about the friends they make in synthetic worlds. Similar to the ties that we build with family members or close friends, comrades in the virtual world can spend a great deal of time with each other. The significant interactions in spaces like *WoW* can even lead to connections so powerful that people develop strong emotional attachments. Curtis explained that he has met people in *WoW* who show definite signs of attachment to the game and the people they meet.

"They are logged in all of the time and always expect you to be available in the game. They even express frustration when you logout," Curtis said. Stories about participation in fantasy-based virtual worlds leading to intimate contact is more common than you might think. Two people who grow fond of each other in a virtual world start exchanging e-mail. This may soon lead to telephone conversations, eventually a face-to-face meeting, and in some instances a serious emotional connection.

Similar to physical-world ties, virtual-world bonds are often temporary. Participation in guilds, for example, are typically fluid, with some users moving in and out of these task-oriented player associations. Curtis explained that he tends to stick with guilds for a certain period of time. When he believes a guild's value has been maximized, that usually means it is time to move on to another guild, another set of challenges, and, importantly, another set of relationships. Speaking about his decision to leave his last guild, Curtis said, "It was quite sad to part with them because it was a lot of fun and they were such good friends." Curtis played with these friends for many hours a day, came to know or trust some of them, but may likely never interact with them again. I call these kinds of virtual-world connections "temporary ties." As long as they last, virtual relationships can involve notable degrees of trust, reciprocity, and interaction.

My point is straightforward. Virtual-world connections are complex social relationships made up of strong, weak, and temporary ties. On the one hand, these affiliations can be strong enough to help synthetic-world users accomplish challenging game-world tasks or, as I discuss below, get through a turbulent period in their lives. Weak ties, on the other hand, may involve infrequent interaction or a brief alliance between synthetic-world users.

Over the course of the social Web's relatively short history, people have developed ties in virtual worlds that they consider just as strong and valid as the relationships they maintain in the physical world. In his study of relationships in MMORPGs, Yee reports that 39 percent of males and 53 percent of females believe that their in-game friends are comparable or better than their physical-world friends. Among the gamers we spoke with, we found similar attitudes.

"You play with these guys for hours every day, so yeah, you develop a bond with them," said Mike, a twenty-four-year-old morning radio intern.

These sentiments will only deepen as the design and technology supporting the use of virtual worlds becomes more advanced. A recent example of this is the arrival of Voice over Internet Protocol (VoIP), a technology that allows synthetic-world users to talk to each other in real time while playing together.

. . .

For the last year and a half, Keith, a twenty-two-year-old gaming enthu-siast, has logged into *WoW* just about every day of the week seeking fun and company with his online friends. Keith told us that he has met hun-dreds of people through his involvement in online games. "I don't think there is any doubt that these guys are my friends," he said. When we met Keith he had recently started using a headset and, like a number of people, credits the device for adding a whole new dimension to his gaming experience. Keith loves the fact that he and his online gaming partners can talk in real time with each other. The voice feature makes navigating the synthetic world more efficient and more enticing.

As you might expect, Curtis also uses voice technology. Like other virtual-world users, Curtis is serious about his reputation and the people he interacts with in *WoW.* Gaming among users like Curtis is more than leisure; it is a very serious endeavor that requires forming strategic alli-ances and relationships. In cases like these, play actually becomes work. At the time of our conversations with Curtis, he was in a raiding guild that was focused on conquering some high-end content.

"I like this guild because it gives me the chance to gain some very valuable items I would never be able to get to on my own," he said. This team of players required elaborate strategizing, the assignment of spe-cific roles, and flawless communication.

Speaking about guilds and the use of VoIP, Curtis told me that "bat-tles are often complex and require a lot of timing and strategy. You don't have a lot of time to move your eyes down for a little text box to read what everyone is saying." Voice, Curtis maintains, makes it much eas-ier to communicate and execute your mission. "Suppose you have a lot of people working together for a common goal," Curtis said. "Let's say there is a big boss that you have to work together to kill."

The ability to communicate in real time makes developing a strat-egy more manageable. But VoIP technology has more than a strategic value. Many gamers believe that the ability to talk with others in-world has a social value too, namely, the ability to strike a more intimate con-nection with other gamers. Keith believes that the use of the headset makes it possible to not only communicate more effectively, but also to bond more genuinely with those he plays online games with. "I talk to people from France, England, Canada, and Mexico every day when I play *Halo* or *Gears of War*," Keith said with assurance.

According to a 2007 study published in the journal *Human Communication Research*, the shift from text-based communication to voice-based communication is transforming gaming—both the way gamers play and how they interact with each other. Drawing their insights from the results of a controlled field experiment, the authors of the study, Dmitri Williams, Scott Caplan, and Li Xiong, found that the use of voice in virtual worlds leads to increase liking and trust among users. Additionally, they found that voice was much more effective than text in dealing with disputes and disruptions among guild members. "Voice," the authors maintain, "was superior for joint task coordination, problem solving, and dealing collectively with dynamic situations." Support for the study's findings surface in our conversations with gamers.

Curtis has met scores of people over the course of his time in *WoW*. We talked at great length with him about the social aspects of *WoW*, the impact of voice, and the kinds of relationships he has developed during his time in the game.

"How many people would you estimate you have met in *WoW*?" I asked.

After pausing for a few seconds to think, Curtis replied, "Maybe around fifty people that I could actually tell you something about."

He interacts with several others through different guilds and estimates that he knows the real-life names of about ten players. These are also individuals that he has collected a bit more personal data about. It is interesting to note that about half of these individuals he met through the game; the other half he knew before interacting with them in *WoW*. His online-only contacts come from all over the world, including Australia and Europe. Bringing off-line contacts and networks into virtual gaming spaces, at the time this book went to press, represented a relatively small percentage of the activity that takes place in virtual worlds. But this is changing and will likely become increasingly common as more people begin to spend greater amounts of time in computer-mediated social places with off-line friends and acquaintances.

■　■　■

In our survey and interviews with young people, they absolutely rejected the idea that the social Web is a place to hang out with their friends. Instead, they consistently expressed a desire to spend time with their friends in off-line third places—bars, coffee shops, and clubs—rather than online third places. Virtual-world enthusiasts, however, are much more likely to view the computer-mediated world as a third place— a location, that is, to gather and hang out with other people. Virtual-world users look forward to going online to experience camaraderie and community typically with people they may never see in the off-line world. Rather than visit a local hang out with friends in the off-line world, heavy users of synthetic worlds are just as likely to visit a computer-generated environment. One *WoW* user we spoke with likened his experience in the game to frequenting that classic third place—a good bar. "You jump into *WoW* and people are already here," he told us. "They aren't always going to be the same people. You just expect people to be there."

Curtis told me that there are days when he logs into *WoW* just to hang out with others. "If I don't feel like doing work in the game, I'll just get together with some friends and we'll go down to the park and sit around and chat." Even in these mostly casual adventures, Curtis and his virtual friends usually stay within character; they continue to wear their role-playing masks.

"What do you talk about?" I asked.

Curtis replied, "Oh, a variety of things. We talk about *WoW*, *WoW* lore, or how people's days have been going."

There is bound to be a certain degree of slippage in moments like these. Consequently, the lines between the online and off-line selves begin to blur.

"I try to keep the two worlds separate," Curtis said.

And yet, he also acknowledged that it is not always easy. Curtis believes his connection to some of the people he talks to regularly via voice has led to greater familiarity and intimacy. He has learned their real names and more about them personally. In his words, "I feel like I can trust them more." Trust, of course, is a hallmark feature of community. Writing about what he calls "social trust"—that is, the feelings of trust humans show toward each other—Robert Putnam observes that "social trust is a valuable community asset." Communities filled with people who trust each other are more efficient, more cooperative, and more engaged with others. Putnam reminds us that "honesty and trust lubricate the inevitable frictions of social life." This is true in both the physical and virtual worlds.

Trust, honesty, and reciprocity are crucial to the making of successful guilds. Like many users of *WoW*, Curtis appreciates the hard work and dedication that goes into building a good, high-end guild. By the time the most skilled players in *WoW* advance to the higher levels, the game becomes less about individual achievement and more about relationships and collaborative play. Given the complexity of some of the raiding guilds he has been a part of, Curtis has learned to trust people he will likely never meet face-to-face—a crucial element of his success. Occasionally, his devotion to *WoW* tests his allegiance to the off-line social networks to which he belongs.

Take the time that Curtis was preparing to participate in a raid that required twenty-five people and such organization that it had to be set up two weeks in advance. He took me back to that particular moment.

"That night I knew at seven o' clock that I was going to log on and get together with twenty-four other people. We were planning on going to this dungeon." He continued, "It was a very complicated raid and we were all relying on each other to carry out their role."

According to Curtis, it is not unusual for the off-line world to intervene and sometimes disrupt even the best-made plans arranged in the virtual world. Balancing allegiances between the physical and virtual worlds can be difficult. That same night a group of his friends invited Curtis to the movies. "Spur of the moment plans," Curtis complained, "are the worst." Looking straight at me, he wondered out loud, "Do I go with the game that I signed up for two weeks in advance [with] people [who] are expecting me to be there? Or, do I go with what I deem to be the better social and better thing to do?" For many virtual-world users like Curtis, the choice is not an easy one to make. In the end, however, they usually chose the game.

Curtis spends more time with his online guild mates than his off-line friends. In his case the virtual social capital he has accumulated is just as viable, if not more so, than the social capital he has accumulated with his off-line friends. This troubled Curtis. Toward the end of one of our conversation he acknowledged that maybe he plays too much. "To tell you the truth, I'm a bit embarrassed by how much I play *WoW*." Most excessive gamers recognize, at some point in their lives, what everyone else does—that heavy use of a virtual world takes you away from the physical world and your off-line friends. When I asked Curtis what the downsides were to his use of *WoW*, he did not hesitate to respond. "I would say the lack of physical activity and interaction with your real-world friends."

Synthetic worlds are so enticing in part because they effectively provide what sociologist Ray Oldenburg calls "spiritual tonic," a place people visit in order to lift their mood and feel good about themselves and the world around them. Similar to the social ties and bonds that are formed in the physical world, virtual-world ties can and often do help synthetic-world users get through the day. And this is a significant feature of what Oldenburg calls "the great good place."

In its 2006 Person of the Year issue, *Time* magazine celebrated "YOU!" In that issue, *Time* proclaimed that Web-based platforms like Wikipedia, MySpace, and YouTube point to a new generation of collaboration and content producers that are "wresting power from the few and helping one another for nothing and how that will not only change the world, but also change the way the world changes." *Time* also hailed virtual worlds as a sign of the changing times. No virtual world generated more buzz than Second Life. In 2007 *Billboard* magazine listed Second Life among the "best bets" that were poised to usher in a new era in media and entertainment. In a reference to Second Life, Bill Werde writes, "The entertainment business has begun to take serious notice, and 2007 should be a breakout year for the virtual world." But the hype about the human migration to 3-D social virtual worlds may be just that—hype.

Overwhelming majorities of the young people we spoke with have never used Second Life. While we have not formally surveyed young people about Second Life, I get the distinct impression that many of

them have very little interest in the 3-D world. Second Life is rarely
if ever mentioned in our interviews or open-ended survey questions.
According to Linden Lab, the company that created Second Life, the
median age of its users is thirty-one. Why are most young twenty-
somethings not experimenting more frequently with Second Life? It
comes down to one simple reason. Second Life is not a reliably social
world.

In a 2008 interview with *VentureBeat*, Mark Kingdon, CEO of Linden Lab, noted that 15 million registered users have signed up for Second Life. And yet, only a fraction of that figure actually uses the 3-D
world. Visit it, and one thing is strikingly clear: nobody is there. Second
Life, it turns out, is not a very sticky world. People sign up, but they
tend not to come back.

Unlike *WoW*, Second Life is not a game. There are no raids or bat-
tles, points to be earned, specific missions to be accomplished, or efforts
to attain a higher skill level. Rather, it is a computer-generated world
that people simply inhabit much like they would the physical world.
People go to Second Life to make all sorts of things—friends, love,
money, elaborate islands, and, most important, a second life. Chase, like
a number of young people we spoke with, is simply not that impressed
with Second Life.

"The name says it all," he said. "All you are really doing is living a
life that is not your own." Discussing how his life and the migration to
digital—social gaming, Facebooking, and keeping up with his friends—
is just about all he can handle, Chase aptly summed up a common view
many young people express about Second Life: "I already don't have
enough time for my own life. I know I don't have time for two lives."

Massively multiplayer online game (MMOG) users consistently list
meeting other people as a favorite activity. But at the time this book
went to press, virtual worlds, in general, are seldom places you would
visit to experience the kinds of things the young people we met tend
to prefer, such as engagement with their off-line lives, interests, and
friends. And this is the second factor that limits young people's interest
in platforms like Second Life. Most of the social interaction in MMOGs
takes place between people who often have little, if any, connection to
each other in the off-line world. This feature violates the main reasons

young people go online—to share their lives with people they know, trust, and care about.

As I explain in chapter 3, young adults consistently reject the idea of the Web as a "third place," a scene for meeting new people. Furthermore, the typical young person we met consistently rejected the idea that online-only relationships are just as fulfilling as off-line relationships. Nearly seven in ten, or 68 percent, of our survey respondents disagreed with the idea that you can get to know someone better online than off-line. Heavy users of virtual worlds differ in this respect. Unlike the majority of young people who spend the bulk of their time online on social sites like Facebook, synthetic-world users are much more likely to believe that online relationships can be just as fulfilling as off-line relationships.

As of now, we simply do not see any evidence that young twenty-somethings are lining up to participate in virtual worlds that function, primarily, as gathering places rather than gaming spaces. At its core, virtual worlds designed for hanging out rather than participating in more collaborative forms of play are simply not the kinds of spaces that draw young people. So, for all the talk about the digital natives, young people's lukewarm response to computer-generated fantasy worlds underscores a key point about their migration to the digital world: their enthusiasm for the social Web is less about the technology and more about connecting to the people that they know, love, and trust. For the majority of young people, the computer-mediated world is about being with real people rather than virtual personas, friends rather than strangers.

SIX

Hooked

Rethinking the Internet
Addiction Debate

> I am addicted to EQ [*EverQuest*] and I hate it and myself for
> it. When I play, I sit down and play for a minimum of twelve
> hours at a time, and I inevitably feel guilty about it, thinking
> there are a large number of things I should be doing instead,
> like reading or furthering my education or pursuing my career.
> —twenty-six-year-old male online gamer

Up until this point I have deliberately avoided using one particular word
in this book—*addiction*. When we asked young twenty-somethings to
describe life in the digital age, the word *addiction* came up frequently.
Users of social-network sites like Facebook and MySpace consistently
told us that these digital destinations are impossible to resist. "Oh my
God, I check Facebook at least ten times a day," said Debra, a nineteen-
year-old psychology major. "I have to admit that even when I'm in class,
I have Facebook open. It is totally addicting." Similarly, in our conversations with users of MMORPGs, the specter of addiction is never too
far away. Many users of virtual worlds believe they have either experienced or witnessed firsthand how the intense compulsion to play games
can lead to serious personal disruptions. "I definitely feel like I was addicted to *World of Warcraft*," James said. "When I first started playing
the game, it was so fun and exciting that I basically gave up everything
and everyone around me."

I often got the sense that the individuals who talked to us, especially
users of social-network sites, were describing behavior that is more habitual than compulsive. Though social and mobile media are relatively

new inventions, few of us can imagine going through life without them today. Like the cars that we drive, communication technologies are necessities that help us efficiently manage the details, small and large, of everyday life. In today's mobile media world we are constantly interacting with a screen or some other device primarily because we can. Have you ever stopped to notice the assortment of people using a communication technology while you are sitting in an airport terminal? Seated across from you is a business traveler checking her e-mail on a BlackBerry. Spread out along a nearby wall are several women from a college volleyball team using their mobile phones to send and receive text messages. And sitting right beside you is a twelve-year-old boy who has powered up his Nintendo DS to play a quick game of *Madden NFL.* This is the digital lifestyle in action.

Meanwhile, addiction implies something altogether different and far more serious—a mental disorder that makes self-destructive behavior nearly impossible to stop.

Jerald J. Block, a psychiatrist at the Oregon Health & Science University in Portland, believes that Internet addiction is a valid mental disorder. Many of Block's patients suffer from what he calls "pathological computer use." In some cases they are addicted to porn. In other instances they may be addicted to a massively multiplayer online role-playing game (MMORPG). Since he began treating compulsive computer users, Block has heard it all. He has treated patients who could not pull themselves away from their computers long enough to take a shower. Others have lost their jobs or friends from the overuse of a virtual world. Some of Block's patients even admit setting up toilets near their computers so that they can remain close to their online friends. Block maintains that the online gamers are harder to treat than other problematic Internet users.

"People feel a lot of shame around computer games," Block says. Ultimately, Block argues that when you spend as many as fifteen hours a day on your computer, it becomes more than work or play. "The computer becomes a significant other, a relationship," he says.

In a 2007 editorial that appears in the *Journal of American Psychiatry,* the Oregon-based doctor encourages his colleagues to include Internet addiction disorder (IAD) in the *Diagnostic and Statistical Manual of Men-*

tal Disorders, Fourth Edition (*DSM-IV*), the guide used by the American Psychiatric Association (APA) in diagnosing mental disorders. Block believes, because of the shame that excessive users experience, as well as minimization by the medical community, that America lags far behind other countries like China and South Korea in dealing squarely with what he calls "pathological computer use." He adds that the issue is further complicated by comorbidity—the existence of multiple, independent medical conditions within patients. According to Block, "About 86% of Internet addiction cases have some other *DSM-IV* diagnosis present." He believes that unless doctors are trained to look especially for problematic Internet use, it will continue to go largely undiagnosed and untreated. Block is not the first to urge the inclusion of Internet addiction in the *DSM-IV*.

As far back as 1995—the year that computers and the Internet began entering more American homes—there has been a surging suspicion that a growing number of people use the Internet not because they want to, but rather because they have to. People like these suffer a truly modern affliction, what is variously called pathological or problematic Internet use. IAD, some maintain, is a genuine diagnosis. Now that being digital is more ordinary than extraordinary, the debate about excessive Internet and new media use is even more urgent.

The idea of computer addiction, according to Internet lore, first gained momentum in 1995 when Dr. Ivan Goldberg, a New York psychiatrist located in the city's Upper East Side, decided to have a little fun with a cyberclub he started in 1986 for his fellow psychiatrists called PsyCom .Net. The number of dependency disorders that are included in the *DSM-IV* astonished Goldberg. A provocateur at heart, he looked up compulsive gambling disorder in the *DSM-IV* and substituted Internet use for gambling. At the time, Goldberg thought the idea of Internet addiction was laughable. If anything, he thought, the widening list of addictions was causing its very own obsession—the need to characterize virtually all of human behavior as addictive. What happened next undoubtedly surprised Goldberg. A number of PsyCom.Net users e-mailed him back to indicate that they suffered from IAD and wanted to know what treatments were available. What started as a joke soon became a matter of serious debate among medical professionals, social scientists, and shortly thereafter, the media and general public.

Throughout the remainder of the 1990s a flood of scholarly papers were published in an effort to establish IAD as a viable sphere of research, medical intervention, and public discourse. Researchers used a variety of methods, including pencil and paper surveys, online surveys, and telephone interviews, to measure Internet addiction. They also created checklists and scales to identify the clinical symptoms and distinct profiles of problematic Internet use. Some of the questions in these scales included: "Have you jeopardized or risked the loss of significant relationship, job, education or career opportunity because of the Internet?" Or, "Do you feel the need to use the Internet with increasing amounts of time in order to achieve satisfaction?"

One of the first and most consistently used metrics to determine Internet addiction is the amount of time spent online. Dependent users of all stripes, the research data suggests, have one thing in common: they spend significantly more time online than nondependent Internet users. Some of the earliest case studies report that the heaviest users spend up to forty hours a week or the equivalent of a full-time job online.

As the methods and approaches to studying excessive Internet use evolved, so did the findings. Take, for example, the gender aspects of excessive Internet use. Many of the early foundational investigations found that men were more likely to display symptoms of Internet addiction than women. Men tended to get hooked on games and pornography. Those claims can be partially attributed to the fact that during the initial rise of the Internet men were more likely to be computer users than women. But as computers and the Internet became more commonplace in the home and women more commonplace in the workplace, the gender gap in computer use essentially disappeared.

Some of the problems researchers associated with excessive Internet use include failure to manage time, a loss of sleep, skipped meals, social isolation, and poor performance at school or work. These, of course, are some of the same problems linked to the abuse of drugs and alcohol. One of the results of the outpouring of research on Internet addiction disorder is the creation of addiction clinics designed specifically to offer

treatment to those who cannot pull themselves away from their computers.

Whether Goldberg liked it or not, a new disorder and domain of
research and clinical practice, Internet addiction, emerged.

In 1996 Goldberg acknowledged that the Internet posed some
challenges. Nevertheless, he insisted that the efforts to classify excessive Internet use as a medically recognized disorder were unwarranted.
"I.A.D.," Goldberg explained, "is a very unfortunate term. It makes it
sound as if one were dealing with heroin, a truly addicting substance
that can alter almost every cell in the body." The doctor the *New Yorker*
described as bearded and burly added, "To medicalize every behavior by
putting it into psychiatric nomenclature is ridiculous. If you expand the
concept of addiction to include everything people can overdo, then you
must talk about people being addicted to books, addicted to jogging,
addicted to other people."

What Goldberg dismisses as ridiculous—the expansive definition of
addiction—others in the medical community are thoroughly embracing
as they seek to learn more about compulsive behaviors and, along the
way, redefine how we think about and subsequently treat them.

Defining addiction is certainly no easy task. Even among the medical
community there is very little consensus when it comes to discussing
addictive behavior in definitive terms. Search for the word *addiction* in
the *DSM-IV* and you will not find it. Instead, terms like *substance dependence* and *substance abuse* appear in the manual.

Here is how the American Psychiatric Association glossary defines
addiction: "Dependence on a chemical substance to the extent that a
physiological and psychological need is established." This, of course, is
the classic definition, the idea that the ingestion of a substance—namely
alcohol or drugs—is an elemental part of any addiction. But a new generation of research and thinking about addiction inspired in part by
fresh discoveries related to the human brain provokes new questions
about compulsive human behavior. In research laboratories across the
world, scientists are looking beyond "chemical addictions" to also consider what are referred to as "natural" or "behavioral" addictions; things

like excessive gambling, eating, or sex. What are doctors learning from the new research?

Doctors have known for years that the ingestion of chemical substances into the human body triggers intense biological reactions. But the details of those reactions are becoming clearer. More precisely, researchers are now focusing on the brain reward pathways, an extraordinarily complex region of the brain made up of a series of integrated circuits that function as the emotional center of the brain, the gateway to feeling things like pleasure, motivation, and reward.

Eric Nestler, a psychiatric specialist at the University of Texas Southwestern Medical Center in Dallas, studies how chemical changes in the brain lead to addiction. Nestler believes that over the course of its evolution, the brain has been wired to capture rewarding experiences in such a way that our body desires them again. Discussing the reward region, Dr. Nestler writes, "It also tells the memory centers in the brain to pay particular attention to all features of that rewarding experience, so it can be repeated in the future."

According to Nestler, this region of the brain is sensitive to environmental stimuli—natural or unnatural. The abuse of drugs over time, for example, can wield a biological influence so intense that it can change the chemical and structural forms of our brain reward pathways in such a way that makes stopping certain behaviors difficult. In extreme cases like these, people continue to take drugs not because they want to but because they have to. The overhaul of their reward pathways compels them to do things that may, in fact, be harmful to their own physical and mental health. Armed with this new body of evidence, doctors now believe that there is a neurobiological dimension to addiction.

The findings related to the brain reward region provokes an interesting question: does the overconsumption of natural rewards—that is to say, non-substance-related rewards such as gambling or gaming—trigger activity in the brain similar to substance-based stimuli? Brain specialists maintain that the pleasures our bodies experience, natural or unnatural, act on the same circuitry system in the brain. Behaviors that the brain finds pleasurable release the chemical dopamine in our bodies, triggering a powerful process that mediates the rewarding effects of environmental stimulus—natural or unnatural. Researchers like Dr. Nestler believe that the brain reward pathways can indeed be chemically and structurally altered by excessive exposure to natural awards.

In a 2008 experiment, researchers from the Stanford University School of Medicine asked eleven young women and eleven young men to play a simple video game. Both the men and the women figured the game out and appeared to enjoy doing so. When researchers looked at the functional magnetic resonance imaging, or fMRI, the brain imaging data repeatedly showed that men felt a more intense feeling of reward from the game than women. In a paper that appears in the *Journal of Psychiatric Research*, the Stanford investigators write that "these gender differences may help explain why males are more attracted to, and more likely to become 'hooked' on video games than females."

It is easy to imagine how online games can trigger powerful activity in the reward pathways of the brain. We know from previous studies that games can produce physiological arousal, including, for instance, increased heart rate and blood pressure. Additionally, it makes sense that epic virtual worlds like *World of Warcraft* can be psychologically tantalizing for millions of users. Complete with fantastic quests, persistent worlds, real-time social interactions, intense battles, and the chance to build a whole new self, virtual worlds provide substantial opportunity for pleasure and reward. Brain research may one day show that the enhanced but computer-enabled sensation of power, influence, status, and control experienced in virtual environments can trigger chemical reactions in the brain reward pathways that make massively multiplayer online worlds alluring, and in the case of some, simply irresistible. For now, though, all of this remains mostly speculative until there is more clinical evidence on the relationship between gaming and the brain's reward region.

This is precisely why in June 2007 the American Psychiatric Association (APA) decided against a recommendation by the American Medical Association (AMA) to officially include games addiction in the *DSM-IV*. Still, the APA's debate that summer set the stage for a new era in research and public discourse about what our relentless engagement with digital technologies may be doing to the human brain and consequently to human behavior.

∎ ∎ ∎

For many people the idea that we could become as addicted to the Internet as to drugs or alcohol is downright silly. What's more, because few randomized controlled trials have been conducted, the empirical evidence some believe is necessary to confirm Internet addiction is scarce at best. Critics claim that problematic use of the Internet is not a disease but rather a symptom of other emotional and mental disorders. For instance: A young man spends excessive amounts of time online looking for excitement and rewards because he suffers from a confirmed disorder like depression. Similarly, a young woman develops an attachment to strangers online because she suffers from loneliness and low self-esteem off-line.

In a 2007 policy report prepared by the AMA's Council on Science and Public Health, heavy video-game use is defined as playing games more than two hours a day. The open-ended and free-form structure of virtual worlds leads the AMA to write in its policy statement that "video game overuse is most commonly seen among MMORPG players." Read the AMA's deliberations in the policy report carefully and it appears as though the nation's most powerful medical body is poised to characterize excessive game use as a mental illness. Despite the leanings of some AMA members, definite consensus regarding games addiction has not been established.

During the AMA's 2007 annual meetings a group of medical researchers decided that more empirical evidence was necessary before listing games addiction in the *DSM-IV*. The APA explained its decision this way: "The APA does not consider 'video game addiction' to be a mental disorder at this time." However, they did leave the door open for 2012, the next year that revisions to the directory were scheduled. "Revising DSM," the APA concludes, "requires a years-long, rigorous process—one that is transparent and open to suggestions from our colleagues in the medical and mental health communities and the public."

Even though the scientific debate about Internet addiction remains foggy, one thing is strikingly clear: both substance-based addictions and natural addictions can dominate people's lives. Block believes that excessive Internet use can lead to a wide array of problems, including "arguments, lying, poor achievement, social isolation, and fatigue." Just as spouses, family, and friends of substance abusers have created support groups to ease their emotional pain and suffering, the companions of excessive gamers maintain their own distinct ways of coping too. Take *EverQuest* Widows, a community that says it is "a forum for partners, family, and friends of people who play *EverQuest* compulsively." Group leaders explain, "We turn to each other because it's no fun talking to the back of someone's head while they're retrieving their corpse or 'telling' with their guild-mates as you speak."

Maressa Hecht Orzack, a clinical psychologist and founder and coordinator of Computer Addiction Services at McLean Hospital in Belmont, Massachusetts, traces games addiction back to the late 1990s. Devices that were once luxuries, the personal computer or a gaming console, gradually grew into commonplace technologies in the steadily rising number of wired castles across America. "Video games," Orzack explains, "used to be contained in arcades, so there were certain limits imposed on the amount of time that you could play them." But with the rise of computers, the Internet, and broadband, games are now available in the home. And with the invention of mobile-gaming platforms, games are available anytime and anywhere. Even if it turns out that games addiction is in fact a mental disorder, the successful marketing and selling of game-based platforms suggests that there are important industry strategies and consumer behaviors that must be considered in any conversation about excessive game play.

Games addiction is not something that only behavioral scientists and doctors talk about; it is a fairly common topic of conversation among gamers and those close to them. In one national survey, 50 percent of respondents, when asked a direct yes or no question, considered themselves addicted to an MMORPG. Many of the young *WoW* users we conducted in-depth interviews with acknowledge that playing the game incessantly interferes with their social lives and personal relationships. Some of the young women we met complained that their boyfriends express more interest in games than in their relationship. Darlene, a twenty-two-year-old college junior, summed it up best: "He spends all of his free time playing that fantasy game [*WoW*]. We rarely go out or do anything together because of his need to get his game fix."

Her language, and that of others we met, is strikingly similar to how a drug addict or alcoholic might be described. In the end the language reflects the widespread belief among young people that addiction to games, specifically, and the overuse of the Internet, more generally, are real issues confronting their generation.

America is not the only nation concerned about the addiction to games, computers, and the Internet. In South Korea, a country where more than 70 percent of the population connects to amazingly fast and afford-able broadband, anxiety about the overuse of games runs extraordinarily high. South Korea is infamous for the thousands of Internet parlors—locals call them PC Bangs—that spread across the country and offer twenty-four-hour access to the Web. In 1998 South Korea hosted an estimated 3,000 PC Bangs. By 2002 the number of PC Bangs had grown to 22,500. In South Korea, PC parlors are the domain of young people. A generation of South Korean youth, primarily males, is coming of age in a cultural milieu in which online gaming is a source of celebrity, cultural ritual, and, according to some critics, a daily obsession.

South Korea is a great case study for better understanding what can happen in a country when a generation of young people aggressively adopts the Internet to the near exclusion of other social and leisure activities. During a 2007 international symposium on Internet addiction hosted in Seoul, it was reported that the average South Korean high school student spends about twenty-three hours a week gaming. The South Korean government estimates that 2 percent of South Korean youth, ages six to nineteen, suffer from Internet addiction and may require some form of intervention—medication, or in extreme cases, hospitalization.

In October 2002 South Korea's Internet dilemma made international headlines after Kim Kyung-Jae, a twenty-four-year-old PC Banger, allegedly died from exhaustion after playing an online game for more than eighty-six hours straight. One detective investigating the case told the BBC that Kim "briefly stopped to smoke cigarettes and use the toilet." In 2006, ten South Koreans reportedly died from blood clots suffered while sitting for extended periods playing games in a local PC Bang. These tragedies, admittedly sensational, underscore what South

Korean officials believe has become a serious public health problem in one of the most wired nations on the planet—a generation of youth who spend sizable chunks of their day sitting in front of a computer screen.

South Korean officials link the allure of computer-mediated leisure and the widespread availability of PC Bangs to a host of social problems. Lee Sujin, a South Korean psychologist who studies Internet use, says "youngsters who become obsessed by the Internet have [experienced] failure at school. They have less interaction with their family and friends and get lonelier." The South Korean government has declared war on Internet addiction by training counselors and creating treatment centers. In 2007 South Korea also created what amounts to an Internet rehab boot camp, believed to be one of the world's first. Jump Up Internet Rescue School's primary mission is to encourage young South Koreans to put down their keyboards and turn off their computers. As one counselor from the boot camp told the *New York Times* in 2007, "It is most important to provide them experience of a lifestyle without the Internet."

In South Korea, much of the speculation about Internet addiction focuses on the young men who spend their days and nights inside PC Bangs. Move beyond the user statistics and the sensational headlines concerning excessive gaming, and it is clear that there is much that we do not know about the young people who play games to the detriment of their own social, emotional, and physical health. In the case of South Korea, several important questions go largely unasked. Who are these young men and why are they spending so much time in Internet cafés? Is this problem common across society or does it disproportionately involve youth from certain sectors of South Korean life? Could it be that young South Koreans are hanging out in PC Bangs not because they suffer a mental disorder but rather because of a broader societal disorder that gives them very little else to look forward to in their lives? So-called natural addictions are not only psychological and behavioral; they are also societal.

Two researchers, Dal Yong Jin and Florence Chee, pursued answers to these questions in their ethnographic investigation of PC Bangs. They concluded that a complex web of factors explain what some characterize as an overuse of Internet parlors. Some use PC Bangs for social contact—that is, to avoid playing alone at home. Others, Jin and Chee note, spend considerable time in PC Bangs because access to broadband is cheap and convenient in the wired parlors. For many others, the researchers discovered, games are a secondary rather than a primary factor for frequenting PC Bangs.

In an article published in the journal *Popular Communication*, Chee refers to PC Bangs as a third place and goes out of her way not to call the users of PC Bangs addicts. She writes, "The PC bang is the site of numerous significant social interactions." PC Bangs are used to gather with friends. In other instances, young people use PC Bangs to engage in courtship practices. And then there are those who turn to PC Bangs during difficult times. In these instances, Chee writes, "A PC bang also has been known to be a cheap place for shelter in the middle of the night, or within the broader context of an unkind job market, a place for the unemployed to spend the day." In other words, the attraction to PC Bangs tends to be socially motivated.

Rather than medicating or hospitalizing heavy users of online games—a view that is gaining momentum in some professional circles in the United States—an equally viable intervention should include addressing the underlying social conditions implicated in heavy Internet use. What is happening in South Korea and other parts of the globe is certainly eye opening, but it is not strictly a medical matter.

Virtual worlds are free and open spaces. The very design of computer-generated environments encourages users to spend extensive amounts of time in-world exploring, leveling, battling, building, or simply hanging out and chatting with other users. And yet these same reasons that make virtual environments such a compelling experience for some can make them an unpleasant experience for others.

As I note in the introduction, the leisure of choice among the four-pack is games. And yet none of the four-pack were active in MMORPGs during the time we spent with them. Three of them held a subscription to either *WoW* or *EverQuest* at one point in their lives but were no longer actively involved with either game. Character abandonment, it turns out, is not unusual in MMORPGs. In one longitudinal study of *WoW*, it was found that of the 75,314 characters observed in the first week of one month, 46 percent were not observed in the first week of the following month. It turns out that some users love everything about massively multiplayer online role-playing games—the adventure, the chance to meet other gamers, and the challenges associated with leveling. And then there are those who try a game like *WoW* and conclude that it simply requires too much time and effort. They turn instead to less time-intensive online gaming experiences. This was certainly true with Derrick and Chase.

"I started *World of Warcraft* in high school, but decided very early on that I was not willing to commit the kind of time that it takes to really advance in the game," Chase told me during one of our conversations.

Derrick's primary reason for giving up *WoW* was similar. "To be good at *World of Warcraft*, I mean really good, you have to give up hanging out with your friends and doing stuff with them." After pausing for a few seconds, Derrick added, "And I was not willing to do that. You know, spend more time with a game than I did with my friends."

Their decision to leave *WoW* underscores a key point about the online world. While participation in the "possibility spaces" enabled by computer-generated environments is truly exciting, there are consequences too, especially for the lives users maintain off-line. In addition to the neurobiological effects of compulsive behavior, there are important social consequences. In the case of young people, excessive media use has been linked to problems like obesity, poor body image, and decreased school performance.

Derrick was a first-hand witness to a former dorm mate who suffered socially and academically from what Derrick characterized as an obsession with *WoW*.

"The guy played *World of Warcraft* all of the time," Derrick recalled. "Whenever we would ask him if he wanted to go out with us, he always said no because he had to stay in and help his guild mates conquer the next raid or pillage the next town."

Derrick not only witnessed the young man withdraw from his friends; he also noticed that the time his peer spent playing the game was most likely responsible for the poor grades that eventually forced him to withdraw from school.

"I don't think any of us who knew him were surprised to learn that he had academic problems," Derrick said, adding that the young man played the game all night, opting to get his rest during the day when he was supposed to be in class.

Derrick's observations are more than speculation. In one national survey of virtual-world users, nearly 20 percent agreed that their use of MMORPGs had caused them some kind of harm—academic, health, financial, or interpersonal. Those who play the most, one study found, are more likely to feel the worst about their involvement in the game. Fifty percent of those surveyed by researcher Nicholas Yee considered themselves hooked. In other words, once they started playing an MMORPG, they could not stop. One of the most enduring consequences of the human migration to the virtual planet is the seemingly insatiable appetite for the social interactions made possible there.

As I got to know Derrick better, he acknowledged his own fight against the impulse to play games all the time. During one of our conversations I learned about an interesting method the college sophomore was using to establish stricter limits on the amount of time he played games. It all started from a conversation with his dad.

"My dad is convinced that games are a waste of time," Derrick told me. "He doesn't get it."

Fully aware of his son's passion for games, Derrick's father told him that he probably spends more time playing games than he does studying. Derrick disagreed. But deep down he knew that he could strike a better balance between the two. His grades in college the first year were fair, but Derrick admitted they could have been better.

Before the beginning of his second year, Derrick and his dad made a wager to determine which activity—studying or playing games—absorbed more of Derrick's time. If Derrick spent more time studying than playing games, his dad agreed to pay him $250. As part of the wager, Derrick agreed to keep a journal that tracked the time he spent doing both. A few weeks after learning about the journal, I asked him how the bet was going.

"I'm learning a lot about myself and how much I play games," he told me.

Shortly after starting the journal, Derrick began confirming what his dad suspected—games were taking up a lot of his time. On his floor, somebody was always playing a game. It was hard to resist the pull of a good game and the enjoyment it usually produces with his gaming buddies. Derrick acknowledged that he did spend a lot of time playing games, but he also believed that he was spending more time studying as a result of the wager.

"The journal," Derrick said, "helps me keep games in their proper place. I feel like I have a lot more balance in my life."

Derrick was not the only person we met whose gaming behavior impacted their academic performance. Remember Curtis from chapter 5? When we met him, he was logging in some hefty hours in *WoW*—eight to ten a day. Curious about the influence of *WoW* on his grades, I asked him when he found time to study.

"I'm currently not enrolled in classes. I decided to take some time off," he said. Curtis intimated that his life was in a state of transition and that he was trying to figure out what was next for him. For one of the few times during our conversations with him, Curtis's voice, usually buoyant, sounded vulnerable.

I did not ask Curtis if he plays *WoW* so much because he is not in school or that he is not in school because he plays so much. As I digested the notes from our conversations with him, I suspect both factors are likely. Research scientists have long suspected that excessive amounts of time spent in front of a screen—for instance, a television or video game—correlates with low academic achievement. Poor academic outcomes are not a result of technology per se, but rather the choices humans make regarding technology. A student, for example, who spends a lot of time watching television or playing games will likely have less time to devote to studying. In one of its policy statements, the American Academy of Pediatrics writes, "Time spent with media often displaces involvement in creative, active, or social pursuits."

And then there is the other issue: does time spent with media interfere with some of our most intimate social ties, namely the ones with our family? From the moment media technologies entered our homes, families have been adapting to their presence. The computer is no exception, as a growing number of families grapple with what it means to manage more and more of their lives through the Internet. Problematic Internet use, we are learning, leads to a host of unintended consequences for families and their personal lives.

As our attachment to the social Web evolves, we can only hope that the public dialogue about problematic Internet use evolves too. The new discoveries about the brain reward pathways are truly groundbreaking. However, one of the dangers of these discoveries is the likelihood that we will soon treat excessive Internet use strictly as a mental illness. A more complete conversation about the overuse of the Internet must comprehend how the relationship between the biological pathways in the brain and the sociological pathways in everyday life combine to influence human behaviors, compulsive or otherwise. Beyond the molecular machinations in the brain, what else makes the online world so irresistible?

The tragic story of three-year-old Brianna Cordell and her mother, thirty-six-year-old Christina Cordell, offers some additional perspective on the Internet addiction debate.

On August 8, 2003, Brianna was found dead in the front seat of her mother's car. A few days later, Ms. Cordell was charged with manslaughter. The autopsy indicated death by environmental hyperthermia. Brianna, doctors believe, died of a heatstroke. A Springdale, Washington, detective assigned to the case told reporters, "We also believe that on the day in question, Ms. Cordell was playing an Internet game, *EverQuest*, for a period of time exceeding two hours, during which she had no knowledge concerning the whereabouts of her daughter."

It was not, the detective noted, the first time Brianna had been seen playing inside of a vehicle with no adult supervision. The local headlines read, "'Addiction' to computer fantasy game may have led to child's death."

Ms. Cordell's compulsion to game was especially curious given her own embattled history with virtual worlds. She was a member of the group Spouses Against EverQuest, an online forum for those who felt abandoned as a result of their partner's fixation with the game they not so jokingly referred to as *EverCrack*. After finishing her posts on the message boards of the group, Ms. Cordell regularly signed off, "A Survivor of EQ Addiction." The mother of two was a frequent presence in the forum, using it to share her struggles with both the virtual and physical worlds. According to Ms. Cordell, her husband's obsession with *EverQuest* lead to spousal and familial neglect and eventually a divorce. Her ultimate struggle in the end was not with the online fantasy game but rather with an off-line life marred by social, financial, and familial instability.

Along with understanding what is going on in the minds of those who spend several hours a day in the online world, we also need to understand what is going on in their lives. In his book *Synthetic Worlds: The Business and Culture of Online Games*, Edward Castronova suggests shifting the conversation regarding the powerful pull of games in this direction. Castronova, a gaming scholar, writes, "For some people," especially those with children and familial obligations, "Earth is where they really ought to spend their time." He notes, however, that "for others, perhaps the fantasy world is the only decent place available." Rather than treat all virtual-world users as a monolith, Castronova challenges us to consider the social place from which they enter into virtual worlds. Thus, what might on the surface appear to be an instance of addiction to a virtual world could, Castronova contends, be a sign of someone "making an understandable choice."

Ms. Cordell admits that *EverQuest* was not a good place for her and her husband. The young mother's life was in a constant state of turmoil. In one of the many references to her ex-husband, she writes, "He didn't want much to do with the kids or me, and then came EQ. It became his obsession." She resented the time he spent playing the game.

"He freaking raided nightly and daily," she wrote to her online peers in Spouses Against EverQuest. "Hell, had to put a porta potty under his fat ass since he wouldn't move practically."

As the couple's marriage descended into emotional pain, jealousy, and spousal abuse, Ms. Cordell managed to find a silver lining in her husband's obsession with *EverQuest*. "In a way," she writes, "I was glad because he wasn't focused on destroying my life as much anymore."

Whatever her misgivings about *EverQuest*, Ms. Cordell began to play it—a lot.

Her reasons for playing were not neurobiological; they were sociological. She was drawn to the online world not because of the powerful pull in her brain rewards pathway, but because of the tragic pathway she traveled in the physical world. A part of her actually loathed *EverQuest* for what it did to her marriage, her family, and her own self-esteem. Still, *EverQuest* was a place to make, if only virtually, a better life for herself. In the virtual world, Ms. Cordell was able to feel things—love, cooperation, trust—that she rarely experienced in the physical world. If she felt powerless off-line, she experienced the sensation of power and control online. If she felt abandoned by her husband in the physical world, she felt desirable when the men she met in the virtual world dueled for her attention and affection. Her journey into the virtual world is not an isolated one.

Over the last two decades a growing number of humans have established residence in virtual worlds for a variety of reasons. In many cases, curiosity or simple intrigue with managing a virtual self or second life is enough to crossover into the computer-generated universe. Some go into virtual worlds to practice elaborate forms of identity play. In cases like these, the virtual universe is a stage to express and experience a new self—bigger, braver, smarter, wealthier, better looking. And then there are those who travel to virtual worlds with a more serious goal in mind—to escape a physical world they find depressing, lonely, or, quite simply, unfulfilling.

Ms. Cordell, a working mother struggling with her own personal demons and an abusive husband, offered constant analysis of the compulsion to play online games. Is it driven by an overwhelming sense of despair? Her response: "The players are acting out their fantasies of what they'd like their lives to be or have more control over their situations, and in reality, they can't."

Is the intense attachment to virtual worlds a way to escape a life full of gloom? Many of her posts suggest that this drove her deep plunge into the virtual world. Immersed in a physical world of pain, the troubled mother of two puts the Internet addiction debate in perspective.

"Although EQ may have only been a symptom of a cause, it definitely doesn't lessen it. It amplifies it," she wrote to her online peers. *EverQuest* did not throw Ms. Cordell's life into a tailspin of despair. The day-to-day costs of living life on the economic and emotional edge did.

Despite playing the game day and night, Ms. Cordell repeatedly expressed regrets, offering this philosophical nugget: "I feel sorry that people get wrapped up into this and destroy their relationships. I just hope that one day they will wake up before it's too late and realize they are alone with no real friends and their family is gone."

While it is true that most of the interactions in MMORPGs take place between people who will likely never meet each other face-to-face, there are exceptions. Games researcher Nicholas Yee reports that a quarter, 25.5 percent, of his male survey respondents and just about four in ten, or 39.5 percent, of his female respondents participate in MMORPGs with a family member—a parent, child, or sibling. In our research we came across an interesting tendency: off-line romantic partners playing massively multiplayer online role-playing games together. For at least two decades, men and women have been using computer-mediated communication to find romance, love, and what MIT scholar Sherry Turkle calls "tiny sex." Scholars studying Internet use have long identified sex as a main motivation for going online. In her book *Life on the Screen: Identity in the Age of the Internet*, Turkle contends that the pursuit of virtual sex "is not only common but, for many people, it is the centerpiece of their online experience." From the moment humans began role-playing, exploring, and participating in computer-mediated communities, sex has been an established fact of online life.

But the journey of romantic partners into the virtual universe is different. In the cases we encountered, couples were not looking for virtual sexual encounters. Rather, they were looking for social interactions with each other. In his survey, Yee found that roughly 16 percent of men and 60 percent of his female respondents participate in virtual worlds with a romantic partner. In some cases the decision by a woman to play *WoW* with her husband or boyfriend is a way of maintaining a connection in the face of the virtual world's demand on her partner's time, energy, and passions. Most women, though, play games for the same reason most men play—they enjoy the identity experimentation and social exploration made possible by virtual worlds.

The 2006 Active Gamer Benchmark Study by Nielsen Entertainment found that more than half, 56 percent, of the nation's 117 million active gamers play online. Sixty-four percent of those, Nielsen reports, are women. According to the report, women play mostly casual online games. But industry reports also show that a small yet noteworthy percentage of women devote their time and energy to the often serious business of collaborative play found in MMORPGs. Though the data varies slightly from study to study, most observers believe that women represent about 20 to 30 percent of the subscriber base to massively multiplayer online role-playing games.

Couple's exploring virtual worlds together is another example of how human behavior is evolving right along with the social and mobile media lifestyle. For those young men and women who find computer-mediated worlds a routine and rewarding source of leisure, the idea of spending time in a virtual world with a romantic partner may one day take its place right alongside other courtship rituals such as dining out or going to the movies. Talking about the joy of playing a MMORPG with her husband, one young woman said, "Overall it can be a cheaper form of entertainment where you can spend quite a bit of time with a significant other." Our discussion with Lisa, a wife in her middle twenties, offers further insight into the reasons why some couples are electing to jump into virtual worlds together.

Lisa was no stranger to games when she and her husband began playing *WoW* together. "I've been playing games most of my life," she told me. She and her husband had been playing *WoW*, in her words, "on and off" for nearly three years when we spoke with her.

"Why did you start playing together?" I asked.

"We like playing games together, particularly cooperative games," she told me. "But, there aren't that many games with a two-player cooperative option. He had already been playing *World of Warcraft* and thought we'd enjoy it."

Initially, Lisa did not think that playing *WoW* with her husband would be fun. But soon after they started she realized how enjoyable it was. Unlike some couples, Lisa and her husband did not start playing together for fear that his attraction to the online world was tearing them apart. "We've played games together ever since we first met," Lisa explained. Games are something that they both enjoy and have in common. They play together not out of desperation, but rather in anticipation of the collaborative play that makes their shared experiences online a meaningful part of their relationship. Underscoring this very point, Lisa said, "Gaming, whether *World of Warcraft* or another game, is a fun activity we can do together, so it's mainly a form of recreation."

But how, I wondered, does playing with a romantic partner work? "Do you log on together, go on quests together, join the same guilds?" I asked.

When they play *WoW* together, they log on at the same time and go on quests in a group. Computer-generated worlds like *WoW* are expansive, even majestic in scope and offer users a variety of people and places to experience. Lisa added, "We each have a Horde character we use for this purpose, kept at the same level. Our Horde characters belong to the same guild. Sometimes, if I'm not available, he'll play on an Alliance character in a different guild." Lisa and her husband do not compete against each other in *WoW*. "It's also fun to form a group and play cooperatively," she said, "because that requires teamwork, which is also a little different than some of the other activities we enjoy."

Think of some of the typical things that couples do together—watch television, go to the movies, or dine out. While couples can certainly do these things with others, it is certainly not unusual to do them only with each other. *WoW*, Lisa maintained, is a unique form of play that she can share with her husband and other *WoW* users who visit the epic world known as Azeroth to have fun, live out their fantasies, and gather among others.

Lisa and her husband are fortunate. Rather than tearing them apart, their mutual attachment to the online world brings them together physically, socially, and emotionally. There is mounting evidence that in many American households a serious challenge to the maintenance of personal and intimate relationships occurs when family members choose to spend more time online than with each other. Dr. Block believes that the intense attachment to the Internet can undermine the quality of our most personal relationships. Talking about the Internet and excessive users, Block says, "First it [the computer] becomes a significant other to them. Second, they exhaust emotions that they could experience in the real world on the computer, through any number of mechanisms: e-mailing, gaming, porn. Third, computer use occupies a tremendous amount of time in their life."

For Ms. Cordell and her husband, *EverQuest* became a significant other. Both of them also developed an attachment to users who presented themselves as members of the opposite sex in the game. Seemingly scoffing at people who develop an unhealthy dependence on the online world, Ms. Cordell proudly insists at one point, "I still have a foot in reality, and school WILL come first before the game. People just need to keep the real world in perspective and not forget what is the real reason to live."

In post after post, Ms. Cordell tried to convince herself that she had control over the game and, more important, her life, telling Spouses Against EverQuest users, "I'm going back to school in over a month to get my degree and become a database administrator and support my two kids since I can't count on a man to do it."

Ms. Cordell's experience with the computer-generated world gave me greater insight into the young people we were meeting like Curtis. It also helped me to better understand the young PC Bangers in South Korea. Brain researchers might argue that Curtis, for instance, plays *WoW* eight to ten hours a day because he suffers from a brain reward pathway overhaul that makes this epic synthetic world too difficult to resist. But as I consider how much time young men like Curtis spend online, I also realize that they play from a particular place in their life.

As our conversations with Curtis evolved, it was clear that he was going through a period of personal turbulence that, in all likelihood, made *WoW* an especially important place for him socially and emotionally. It is quite plausible that Curtis plays *WoW* because the game enables him to experience things online—achievement, excitement, and pleasure—that he has yet to adequately experience off-line. In other words, he plays as much for social and personal reasons as he does for neurobiological reasons.

For all of its promise and potential, the fact that some people, a few million perhaps, go to virtual worlds to feel things—power, love, status, self-worth, and acceptance—that they cannot feel in the physical world is distressing. In the end, this suggests that the reasons people use computer-generated worlds may have as much to do with what they encounter in their first lives and physical world as it does in their second lives and virtual world. Even as we learn more about the biological aspects of the compulsion to go online, we must never overlook the social aspects. Life in the online world is intricately connected to life in the off-line world. It always has been and it always will be.

SEVEN

Now!

Fast Entertainment and Multitasking in an Always-On World

> For the younger generation of multitaskers, the great
> electronic din is an expected part of everyday life. And given
> what neuroscience and anecdotal evidence have shown us, this
> state of constant intentional self-distraction could well be of
> profound detriment to individual and cultural well-being.
> —Christine Rosen, senior editor of *The New Atlantis*

Laptop computers, mobile phones, and iPods deliver a vast assortment of digital media content in unprecedented speed—now! Today, we can view video clips, listen to our favorite music downloads, squeeze in a game, access user-generated content, or just about anything else in smaller and quicker doses, thus making these digital delights all-pervading and irresistible. A 2007 *Wired* magazine cover story titled "Snack Culture" celebrates the rise of what it calls bite-size entertainment and the emergent world of one-minute media. "We now devour our pop culture," writes Nancy Miller, "the same way we enjoy candy and chips—in conveniently packaged bite-size nuggets made to be munched easily with increased frequency and maximum speed." *Wired* calls it snack culture. I call it fast entertainment—this ever-widening menu of media content that we can consume easily and on the go.

No matter where we are, fast entertainment is generally just a click away. I liken the efficient delivery of digital media content to another staple in the daily American diet: fast food. Like fast food, fast entertainment is easy to get, all around us, and typically cheap, but not always good for you.

It is hard to know which came first—our appetite for speedy entertainment or the widespread manufacturing and aggressive marketing of
it. Mobile technologies are ideal platforms for delivering bite-size media easily and effectively. For much of its early life, fast entertainment
was the primary domain of technology companies, Silicon Valley start-
ups, and Web entrepreneurs. Known for its user-generated content, a
major part of YouTube's success can also be attributed to the creation of
a space that supplies an endless stream of short video clips that comple-
ment our desire to consume a lot very quickly. Yahoo! is also a favorite
online destination to grab a quick video clip, music playlist, or game.
Social-network sites allow users to share a wide range of content with
their fellow users casually and throughout the day. Come across a funny
video and send it to your Facebook friends. Discover a cool new underground hip-hop band and spread the word through MySpace. Nowhere
is the hegemony of fast entertainment more evident than in the rise of
Apple's iTunes Store, one of the most prolific platforms for delivering
digital media content.

In his book *The Perfect Thing: How the iPod Shuffles Commerce,
Culture, and Coolness*, *Newsweek* chief technology correspondent Steven
Levy explains how the music industry insiders dismissed the idea of
selling digital downloads. Apple's CEO, Steve Jobs, told Levy, "When
we first approached the labels, the online music business was a disaster.
Nobody had ever sold a song for ninety-nine cents. Nobody ever really
sold a *song*." When Apple launched the iTunes Store in 2003, the service
developed a reputation for delivering digital music downloads quickly,
reliably, legally, and cheaply. From day one the iTunes Store was a huge
success, reportedly selling 275,000 downloads in its first eighteen hours
of operation. By day five, the number of downloads reached one million.
In six short years, iTunes went from being nonexistent to surpassing the
retail behemoth, Wal-Mart, to become the number one music retailer
in the United States.

In 2005 the iTunes Store began offering downloads of music videos
and popular television programs for the iPod. Soon after that the store
expanded to include movies, podcasts, audiobooks, and games, among
other things. Apple's iPhone has been an even bigger hit and distribu-
tor of fast entertainment. In just nine months iPhone users downloaded

more than one billion applications, many from categories like games, music, and entertainment. Apple's iPod and iPhone made it convenient and cool to carry our entire inventory of media and entertainment in our pocket. No matter if you are at home or on the road, access to your media entertainment library is just a touch of a button away. Technology companies such as YouTube, Yahoo!, and Apple were among the first to superserve our desire to quickly and constantly consume media. After realizing our desire to consume content on the go and in smaller bites, the old media guard is responding.

From the big music labels to the network television industry, the major media companies are investing, strategically and financially, in fast entertainment by offering up their own versions of bite-size media. In 2007 Sony Corporation and its production studio, Sony Pictures Television, tossed its hat in the fast entertainment ring. The studio announced that it would begin sifting through its massive inventory to produce three-to-five-minute versions of programs that once filled the thirty-minute and hour-length programming slots in the primetime schedule. Sony executives call these tiny TV episodes minisodes. According to the studio, minisodes are not clips but rather full episodes containing a discernible narrative structure consisting of a beginning, middle, and end. The initial roster of miniature programs included old television series like *Charlie's Angels*, *T. J. Hooker*, and *Starsky & Hutch*.

Sony executives characterize the minisodes as campy and fun, but the decision to produce them is a serious response to how young people enjoy their media. In many ways, the creation of three-to-five-minute episodes are influenced by a video-clip culture that is a staple in the online experiences of many young people. From Lime Wire to BitTorrent, the adoption of file-sharing platforms not only alters how the young and the digital consume media content; it also reflects a greater interest on their part to manage, share, and engage content.

One of the more intriguing paradoxes of today's digital media environment is that we consume more and less at the same time. How is this possible? The rise of YouTube, founded in February 2005 by Chad Hurley and Steve Chen, is a great illustration. They launched the video file-sharing site to empower users of social media to do with videos what they were doing with photos—create, manage and share them with peers. You name it—music videos, sports highlights, news segments, political speeches, user-generated content, clips from television and film—and YouTube users share it. According to comScore, "Americans viewed more than 11.4 billion videos for a total duration of 558 million hours during the month of July 2008." Five billion, or 44 percent, of the videos were watched on YouTube. More precisely, 91 million viewers watched a YouTube video, which averages out to about fifty-five videos per person. The average length of a video watched in July 2008: 2.9 minutes. In short, Americans watched a ton of online videos, the vast majority of them short clips. In the digital media age, more equals less.

YouTube, in many ways, represents the next evolution in how we consume video—quickly and conveniently. Today we digest bits and pieces of an ever-sprawling narrative universe. This type of watching—call it browsing—began before the arrival of the Web. Since the widespread diffusion of the remote control in 1990, we have grown accustomed to watching bits and pieces of multiple programs at the same time, shifting our allegiance and attention in between commercial breaks, narrative lulls, and a need to sample as much content as we can. YouTube is the ultimate remote control. The platform makes it easier to do what we were already doing—watching a little bit of everything simultaneously. In 2007 the average U.S. home received 118.6 channels, according to Nielsen Media Research. YouTube raises that by infinity.

In addition to consuming less of more, we are constantly consuming. We have evolved from a culture of instant gratification to one of constant gratification. Fast entertainment encourages an insatiable desire to be entertained no matter where we are (at work) or what we are doing (driving a car). While the scientific merits of Internet addiction are still being measured, analyzed, and assessed, one thing is undeniable: the desire for fast entertainment is widespread and voracious. We not only want our media content now, we expect it now. Not that long ago, consuming media on the go was a luxury. Today, it is a standard feature of daily life.

In many ways, the social- and mobile-media lifestyle represents a new cultural ethos and a profound shift in how we consume media—in smaller and steadier portions and on smaller and more mobile screens. Along with changing how we consume media, fast entertainment changes where we consume media. Two developments defined the post-war boom Americans experienced starting around the 1950s: a rapid increase in home ownership and a bustling consumer economy. By the 1950s Americans were furnishing their new homes with all kinds of appliances that helped to establish a more modern and socially mobile lifestyle. Beginning with the radio, the phonograph, and then the television, Americans also filled their homes with media. Over the next half century, Americans participated in a flourishing home-centered media consumer culture that redefined leisure and household life.

Not that long ago we typically left our homes to purchase a wide array of media—music, books, magazines, videocassettes, games, DVDs—and then returned home to enjoy it. Today, however, the reverse is increasingly more common. We collect content from the comfort and convenience of our home, via digital downloads and peer-to-peer networks, to take with us when we leave our home. Throughout the last two decades of the twentieth century, Americans turned their homes into what media technology professor Jorge Schement calls the wired castle. Our homes became the ultimate leisure space. But in an always-on world, any place can be a good place to grab a quick bite of entertainment.

Despite all the euphoria over iTunes, iPods, iPhones, Webisodes, minisodes, minigames, and one-minute media, is the ability to be entertained constantly and no matter where you are really a good thing?

Young people are media rich. They own music players, computers, mobile phones, TVs, and game consoles. Young people's media environment is like a kid who wakes up one day and finds himself in a candy store. Surrounded by so many tasty options, what does he do? Naturally, he devours as much of it as he can, any way that he can. And that is essentially what we are seeing young people do with media. Immersed in a world of media, they use as much of it as they can, any way that they can. Innovative as ever, the one sure way for young people to use all of the media and technology they own is to use it simultaneously. One study, for example, found that American youth report spending about six hours a day with media. If you include the simultaneous use of me-dia, that figure grows to about eight hours a day. Communication schol-ars refer to this as media multitasking.

We all media multitask. As I work on this chapter, my e-mail box, music player, photo application, and several tabs on my two favorite browsers are all open on my computer screen. But the young and the digital are widely viewed as masterful multitaskers, capable of managing several technologies, screens, and conversations fluidly and simultane-ously. They multitask habitually and according to many observers, they also do it instinctively.

Donald F. Roberts, a Stanford University communications profes-sor, has studied the media behaviors of children and adolescents for more than thirty years. Roberts believes that adolescents' multitasking ways began to really take shape as the media in their homes migrated into their bedrooms. With the use of ethnographic techniques, time di-aries, and surveys, researchers began building more nuanced portraits of the media environment in American homes in the 1970s. A 1972 study of southern California sixth graders found that 6 percent of the sample had a television in their bedrooms. Since then, the flow of media into children's bedrooms has continued at a steady clip. In his study of media use by youth, Roberts and his colleagues found that no matter if they were especially young, ages two to seven, or older, ages eleven to four-teen, kids in America access a lot of media from their bedrooms.

At least a third of young children, for example, can watch television or listen to music from their own bedroom. Similarly, three-quarters of older children are able to watch TV or play a video game from the comforts of their bedroom. By the year 2000, about 20 percent of older youth, ages twelve to eighteen, were accessing computers from their bedrooms. Assessing the state of young people's media environment by the start of the new millennium, Roberts writes, "Compared with even a few years ago, the sheer numbers of children and adolescents possessing personal media is remarkable." The movement of media into children's bedrooms creates the context for more frequent, intense, unsupervised, and in the view of some researchers, unhealthy media behaviors.

Until recently the data on media multitasking was extremely limited. How much do young people media multitask? According to a 2006 study by the Kaiser Family Foundation, a lot.

The San Mateo, California–based research unit surveyed a national cross-section of 2,032 school-age kids ages eight to eighteen. In addition, Kaiser analyzed data from a self-selected subsample of 694 respondents who completed a seven-day diary of their media use. Here are some of the study's findings. Children and teenagers spend at least a quarter of their time with multiple media. On a typical day, Kaiser reports, eight in ten school-age children media multitask. Predictably, young people do a lot of their multitasking when they are using a computer. Older children, ages fourteen to eighteen, multitask more, but by the age of ten, their desire to multitask is strong. Young multitaskers tend to use other media—such as television or music—while on the computer. Even when using the computer is the only activity, it is common for young people to shuffle back and forth between instant messaging, their favorite Web sites, online videos, and games, all while managing a wealth of digital content such as photos and music files.

The intense media multitasking behaviors among the young and the digital began forming just as computers designed to handle multiplying windows and applications entered our homes. We hear it often and it is true: young people have not known a world without personal computers. Equally noteworthy is the fact that they have not known a world without computers and software capable of handling multiple tasks at the same time. From their view of the world, the computer has always been what one research group calls a "multitasking hub." They do not think twice about engaging multiple screens (e.g., the television, computer, mobile phone), multiple windows on their computers (e.g., AIM, iTunes, MySpace), multiple tasks (e.g., checking their Facebook news feed while doing homework), or multiple conversations at the same time. Researchers call it multitasking, but young people have another name for it: life.

As we collect young people's technology life histories, media multitasking is a reoccurring theme. Many young twenty-somethings, for example, vividly recall using music media, games, and television and managing multiple conversations via AIM while doing homework. In many instances, all of these platforms were available in the privacy of their own bedrooms where there was little if any parental supervision. For a generation that has grown up with a wide array of media technologies in their homes and in their lives, multitasking is as natural as the air that they breathe.

Tweens and teens are not the only ones multitasking their media. Our surveys and in-depth conversations show that college students are avid users of multiple media too. They multitask with just about every media they use—the Internet, music, television, and books and magazines. A decisive majority, 95 percent, listen to music either most or some of the time when using the computer. Seventy-two percent of our survey respondents said they watch television most or some of the time when using a computer is their primary activity. And when television is the primary activity, 81 percent of our respondents said they use a computer either most or some of the time. That is what Justin, a twenty-one-year-old psychology major, does.

"I multitask the most with the computer and television," Justin explained. When there are commercials he usually goes online. "In that gap of time, I am normally on instant messenger talking to friends."

Twenty-one year-old Andrea, an advertising major, uses television and the Internet at the same time too. "For me, the television works as background noise and the Internet allows me to be connected at all times," she said. When she gets bored with television, her computer is never too far away. Different media require different cognitive loads. For example, the load needed to listen to music is considered less than the cognitive resources needed to read a book. One thing is consistently clear in our research: among college students, using one media almost always means interacting with other media too.

Multitasking media habits are formed relatively early and right around the time young teens begin to develop their own peer networks, media interests, and greater independence from their parents. Among the young people we met, media multitasking is widely accepted as a fact of everyday life. Johnson, a nineteen-year-old, summed up multitasking this way: "I don't really think about it. It's just something that I've always done." Twenty-two-year-old Justine's multitasking skills are at once typical and amazing. "At one time I can be banking, paying bills, checking my e-mail, Facebooking, e-mailing my parents, talking online to my friends, checking the *TV Guide* on the Internet, and researching possible graduate schools," she said. As one young woman put it when referring to her generation, "Multitasking is easy and natural for us."

She's right. Multitasking for the young and the digital is easy and it certainly appears natural. But a growing body of evidence provokes the question, is media multitasking effective? More important, is it healthy?

When we asked young people why they multitask, the response was consistent: to accomplish more things efficiently. Twenty-two-year-old Brandon said, "I would never get anything done if I did not multitask." Most of us, in fact, multitask as a way to more effectively manage our time. And yet, even as humans continue managing multiple screens, media, and tasks simultaneously, cutting-edge brain research is beginning to confirm what some say is obvious: doing several things at once actually reduces task efficiency and proficiency. There is growing evidence that multitasking may not only slow down the completion of tasks but may also impair our performance. Addressing our incessant desire to multitask in a piece that appears in the *Atlantic*, Walter Kirn writes, "The great irony of multitasking—that its overall goal, getting more done in less time, turns out to be chimerical."

In 2007 a team of psychology professors working from the Human Information Processing Laboratory at Vanderbilt University conducted a series of test trials to assess the brain's capacity to perform multiple tasks. For years brain specialists have suspected that the brain contains a bottleneck function that keeps us from concentrating on two different things at once. Dr. René Marois, one of the principal investigators from the Vanderbilt lab, focuses on the neural basis of attention and information processing in the human brain. More precisely, Marois and his colleagues seek to more fully understand why humans appear unable to execute more than one mental task at a time. Using functional magnetic resonance imaging (fMRI) of the brain, Marois and his colleagues identified what doctors believe is the mechanism that prohibits humans from processing two or more things at the same time, the central bottleneck of information processing.

To determine what happens when we ask our brain to execute two tasks at the same time, the Vanderbilt researchers conducted dual-task trials. In the first task participants were instructed to touch the appropriate button in response to a sound stimulus. In the second task participants were asked to select the appropriate vocal response to a visual stimulus. Compared to the complex information processing that takes place daily, each task in the trials is easy to perform. Hear a sound, push a particular button. See an image, utter a specific word. Easy, right? But the Vanderbilt team made things more interesting and challenging by varying the time interval in between these two simple tasks. In some instances participants were given 300 milliseconds in between tasks, while others were allotted more time, on average about 1560 milliseconds.

Among the test subjects with very little time in between the first and second tasks, there was a statistically significant delay in executing the second tasks. The experiments along with the brain scanning data from the fMRI provide neural evidence of what researchers call dual-task interference. That is, when we try and process two pieces of information simultaneously, a traffic jam ensues in the brain. Conversely, among the subjects allotted more time in between each task, there was no significant delay in their execution of the second task. In addition to executing the trial experiments more efficiently, the subjects with a longer interval between tasks were more likely to push the correct button or execute the right vocal response. In short, subjects with more time in between tasks were much more efficient and proficient. Our brains, it turns out, are not wired to process dual information simultaneously. Results of the study were published in the December 2007 issue of the medical journal *Neuron.*

In the summary section of the peer-reviewed article, the Vanderbilt researchers write, "When humans attempt to perform two tasks at once, execution of the first task usually leads to postponement of the second one." The results of the trial experiments, the doctors maintain, "suggest that a neural network of frontal lobe areas acts as a central bottleneck of information processing that severely limits our ability to multitask."

With over one million neurons and a highly complex circuitry system, the brain is a powerful machine. Still, even the human brain, as remarkable as it is, has limitations. Discussing the implications of the dual-task experiments, Marois told the *New York Times*, "We are under the impression that we have this brain that can do more than it often can." Marois goes on to say that "a core limitation is an inability to concentrate on two things at once."

The significance of these findings extend well beyond medical labs and scientific journals. Multitasking has real-life implications both for our brains and our world. Think about how frequently we multitask throughout the day. In some situations multitasking—say, responding to e-mails while eating lunch—may be quite useful in managing a busy day. In other situations multitasking may be inappropriate and even fatal. That was certainly the case on September 20, 2008, when twenty-five people were killed after Metrolink 111 crashed head-on with an oncoming Union Pacific freight train in Chatsworth, California. It was the worst California train disaster in fifty years. An investigation revealed that, seconds before the crash, the Metrolink engineer at the helm of the train was sending and receiving text messages. Concentrating on his phone messages meant that he could not focus on his path and a series of warning signals that would have almost certainly prevented the accident.

At this point and time there are as many questions as there are answers when it comes to understanding the neurological implications of multitasking. For instance, some brain specialists believe that constant multitasking may be stretching our neural capabilities beyond their outer limits and subtly changing the machinations of our brain. Studies show that certain regions of the human brain are wired to process information, whereas other regions facilitate our ability to recall information. There is growing speculation in the medical community that young multitaskers may be conditioning their brains to quickly access, manage, and process information while underdeveloping the neural ability to recall and understand the information that they find.

Popular notions of media multitasking are misleading. The person using a computer while watching television and responding to text messages on their phone actually uses one screen at a time. Multitasking involves switching one's attention back and forth from one platform to another. In reality, the issue media multitasking raises is not simultaneous media use per se, but rather the ability of humans to pay attention in an always-on, always-connected digital-media environment. Underscoring this very issue, Christine Rosen, senior editor of the *New Atlantis*, writes, "When we talk about multitasking, we are really talking about attention: the art of paying attention, the ability to shift our attention, and, more broadly, to exercise judgment about what objects are worthy of our attention."

Now that anytime, anywhere technology and fast entertainment are pervasive parts of our cultural environment, deciding what to pay attention to is more challenging than ever. Linda Stone, a communication technology thought leader and consultant, believes that the constant efforts by humans to manage our time—a main reason we multitask—should be accompanied by an equally zealous effort to better manage our attention. In an age of multiplying screens, constant connections, and content overload, Stone, a former Apple researcher, believes that humans suffer severe lapses in attention. She even has a name for this particular state of being: continuous partial attention, or CPA. Whether or not humans are genuinely addicted to the Web is still a major source of debate, but one thing is undeniable—managing our attention in a world of anytime, anywhere technology is one of the great challenges of modern life.

Stone makes a distinction between multitasking and CPA. With multitasking, she maintains, "we are motivated by a desire to be more productive and more efficient." We multitask to save time. Conversely, Stone writes, "we pay continuous partial attention in an effort NOT TO MISS ANYTHING." She adds that CPA "is an always-on, anywhere, anytime, anyplace behavior that involves an artificial sense of constant crisis. We are always in high alert when we pay continuous partial attention." CPA describes a familiar yet relatively recent state of being—the constantly tethered to technology lifestyle. No matter if it's sending a text message, responding to e-mails, tagging the latest batch of pictures posted by a friend online, or downloading the latest application for your cool new phone, the digital world is a busy world. The nonstop access to content and comrades via smaller and more mobile screens keeps us on constant alert. Meanwhile, our attention stays on the move, constantly shifting from one task to the next, one conversation to the next, one screen to the next.

Whereas CPA is an increasingly normal state of being, it is not a very healthy state of being. As a result of CPA, Stone argues that "in a 24/7, always-on world, continuous partial attention used as our dominant attention mode contributes to a feeling of [being overwhelmed], over-stimulation and to a sense of being unfulfilled."

Along with understanding the neurological implications of multitasking, we need to understand the sociological implications too. Millions of people talking on their phones while driving make the roads less safe. Multitasking while doing homework, a common behavior these days, can contribute to poor academic performance. For robust multitaskers like the young and the digital the stakes are even higher, the outcomes potentially more profound. As Rosen notes, "For the younger generation of multitaskers, the great electronic din" is a common aspect of life. Rosen notes, however, that "when people do their work only in the 'interstices of their mind-wandering,' with crumbs of attention rationed out among many competing tasks, their culture may gain in information, but it will surely weaken in wisdom."

Fast entertainment, media multitasking, and continuous partial attention are especially ominous in environments where focus, attention, and concentration are essential to succeed. Nowhere are these issues more urgent than in what is, arguably, the most important place in the lives of young people: school.

EIGHT

"May I have your attention?"

The Consequences of Anytime, Anywhere Technology

I check my Facebook at least once every hour at work.
—Kayla, twenty-two-year-old college student

Sitting in a neighborhood café one day, I unexpectedly found myself doing some ethnographic work as I witnessed the social- and mobile-media lifestyle up close and personal. Not long after I had powered up my laptop and settled in, a group of seventh and eighth graders started filing into the quiet little shop. A nearby middle school had just let out for the day, and the vibe of the place changed instantly. Both the middle school girls and boys had that just-got-out-of-jail look in their eyes and were boisterous, joking, and generally upbeat. Some of them were hanging out waiting for a parent to pick them up. Others were digging through their pockets and backpacks for money to purchase an after-school snack. As a group of them settled in their seats, they all pulled out cell phones and began checking and sending text messages. A few pictures were snapped with the phones. There were also some quick tutorials about this or that application, the kind of peer-to-peer knowledge sharing that makes the young and the digital such formidable players in the new media landscape. I even noticed an iPhone among the bunch, which prompted me to quietly slide my modest mobile device beyond their view.

What I was seeing, of course—young people socializing with each other face-to-face and through their mobile phones—is standard fare today. Around the world, the phone is a centerpiece technology in young people's lives. In a 2008 study of 2,089 mobile-phone users in the United States, ages thirteen to nineteen, Harris Interactive found that 45 percent agree with the statement "Having a cell phone is the key to my social life." In most families today the question is not if school-age kids will get their own mobile phone but rather when. By most accounts kids are asking for and getting mobile phones at earlier and earlier ages. For them, getting a phone is a life-changing moment that signifies a greater degree of maturity, personal mobility, and independence. By 2007 a decisive majority of American teens, 77 percent owned a mobile phone. Studies show that about 40 percent of tweens, kids ages eight to twelve, own phones.

Anytime, anywhere technologies have trickled down to the youngest members in our culture. Young children can operate DVRs, smartphones, and laptops with great ease and efficiency. Among my friends and peers it is not uncommon for children as young as six to own an iPod. The music purchased by tweens and teens is commonly among the top downloaded at the iTunes Store. In fact, young people have turned the iTunes Store into a virtual mall. In addition to shopping at the Apple site, they excitedly share their views about the pop music and icons marketed to them by corporate music labels.

An unintended consequence of young kids' adoption of digital media is that fast entertainment and continuous partial attention (CPA) are invading our nation's schools.

Today, teachers and school administrators are on the frontlines of the digital migration. Every day at work they face a generation who own more personal and mobile media than any cohort of kids the world has ever known. In order to learn more about the impact in the classroom, I visited schools and spoke with several principals and teachers.

When I initially began these conversations, I was surprised to find that a growing number of schools permit students to bring their personal media to campus. Not that long ago, the idea of allowing students to bring transistor radios or Walkmans into the school was unthinkable. But the current decision to allow kids to walk into schools equipped with a personal army of media reveals how our values, behaviors, and culture are evolving in the digital age. Two factors, ultimately, led to a more open policy regarding the presence of mobile phones in America's classrooms.

First, the open policies acknowledge what Everett M. Rogers calls the "diffusion of innovations." Back in the 1960s, Rogers developed his now widely cited theory to help explain how technological innovations spread throughout society. Innovations are initially used by what Rogers refers to as "early adopters" before gaining widespread and in some cases near-universal embrace across a population. Compared to countries in parts of Asia and Europe, the United States was a laggard when it came to adopting the mobile phone. But a combination of factors, including sharp pricing decreases, competitive family phone plans, and disparate family schedules, have greatly increased mobile-phone use in America. Another factor is the anxiety caused by tragedies like high-profile school shootings and the attacks on September 11, 2001. Just how much has mobile-phone use in the United States evolved?

In a 2002 poll by the Pew Internet & American Life Project, 38 percent said it would be hard to give up their mobile phone. In that survey more or just as many people found it hard to give up their landline phone (63 percent), television (47 percent), or the Internet (38 percent). Five years later, though, the technological preferences of Americans revealed significant change. Mobile phones (57 percent) ranked ahead of the Internet (45 percent), television (43 percent), and the landline phone (40 percent).

Among young people, the mobile phone is an especially desirable technology. Sixty-two percent of young people told Pew it would be hard to give up their phone. According to Pew, Americans under the age thirty "are much more likely to say it would be hard to be without a mobile phone than to be without the internet or email." In 2000 the most likely user of a mobile phone in the United States was a white, middle-aged male professional. By 2008 a broad cross-section of Americans—men and women, young and old, students and professionals—used mobile phones. And when it comes to some of the more creative and personalized aspects of mobile-phone use, individuals under thirty are the trendsetters.

In short, school administrators around the country get it—mobile phones are a fact of life. Ms. Waters, the principal of an affluent high school, explained her school's decision to allow personal media this way. "Rather than fight our students, we have decided that the most effective approach is to help them handle their media at school more responsibly." She added that "devices like cell phones and iPods are way too small to try and police. They can put these miniature gadgets in their backpacks or pockets and we would never know they had them." As long as the devices stay out of sight and are used properly, she insists that the school should focus on other matters. At her school, students are allowed to bring mobile phones, laptops, and iPods. They can use them during lunch and in between classes. Teachers in her school can use their own discretion regarding personal media in the classroom. A vast majority of the teachers that I spoke with believe that the decision to allow personal media in the school is a mistake.

One teacher stated flatly, "Once you start allowing kids to bring cell phones and iPods into the school, you are really asking for trouble."

The second reason schools have decided to permit mobile phones in the school appears to be more out of outside pressure than a genuine embrace of such policies. When principals and teachers initially moved against mobile phones, an unexpected voice of protest rose up against the idea—from parents. Many teachers and principals told me that parents are the strongest advocates for allowing their children to carry mobile phones to school.

"Parents," one principal told me, "absolutely insist that their children carry phones."

Mobile phones are central in the management of household life. Family members often have disparate schedules that render them busy and on the go. Mobile phones are an efficient way to communicate with family, coordinate schedules, and update whereabouts. For many parents the use of a phone by their children is no longer a luxury, but it is a source of security.

The nation's largest school district has managed to resist pressure from parents to buck a national trend that permits mobile phones and personal media in the school. With over 1.1 million students under its watch, the New York City Board of Education moved aggressively to rid its schools of mobile phones, iPods, and other devices officials believe undermine the academic and learning experience. This is part of a larger initiative by the city's mayor, Michael Bloomberg, and School Chancellor, Joel Klein, to improve student school performance by creating what they characterized as a safer and more prolearning environment.

On April 26, 2006, New York City schools began randomly installing metal detectors to search for what it called "school contraband," or objects that the new disciplinary code prohibited students from bringing to school. But the policy to keep banned objects out quickly turned into a crackdown on mobile phones in the schools. Over the first twelve days of the plan, the New York City Department of Education reported capturing seven knives, two box cutters, a razor, some marijuana, and around eight hundred phones. Many New York students, especially those from middle-class homes, were shocked when they arrived at school only to find that the metal detectors were there to conduct random searches. There were local news reports that some students broke down in tears when they were forced to relinquish their most prized possession, their mobile phone. When the school year resumed in September, the first five weeks of sweeps led to the confiscation of 34 weapons, 784 electronic gadgets like the iPod, and 2,286 phones, according to the city's department of education.

In an interview with a local NBC news affiliate, Klein explained that the mobile-phone ban was necessary to maintain order and discipline. "They are used for everything," the chancellor told NBC News. Klein elaborated: "Kids are text messaging in their pockets. Phones can be used for people to be on the Web and, you know, we don't have time during the school day to spend our time chasing around the cell phones. We need to educate our kids."

NYC school officials complain that mobile phones also contribute to the making of poor citizenship and an unhealthy school environment. Some students use their phones to bully, isolate, and intimidate their peers. And with the added bonus of camera phones, school officials note that students use their phones to take and circulate inappropriate pictures of people in the locker room. A few teachers even complained that students were snapping pictures of exams and then passing them on to friends.

Bans on mobile phones in schools date back to the mid-to-late 1980s when fears that students were using beepers, pagers, and phones to sell illegal drugs ran rampant. By the start of the new millennium the public image of the typical teen mobile-phone user had changed dramatically as tweens and teens from middle-class and affluent households turned the phone into a mobile-media platform (i.e., music downloads), a source of personal expression (i.e., customized ring tones), a content creator (i.e., pictures), and a whole new language (i.e., text messaging).

Mobile phones are a way of life for Generation Text. A 2008 study found that mobile phones ranked second only to clothes when it came to determining the social status of teens. In an appearance in front of state lawmakers considering a ban on mobile phones, one young teen defended cell phones this way: "To you this is a tool. To me this is jewelry." For her generation, mobile phones are as necessary as the clothes you wear to school. Reacting to her phone being confiscated, one fifteen-year-old girl cried, "I feel naked. I feel like I lost something very important to me."

The zero-tolerance policy provoked an intense showdown with the parents of New York City school children who called the ban out of touch with the way families are managed today. New York's open school policy means that many students take the subway to get to far-away schools and back home. In their e-mail messages to the school board, parents characterized the policy as "cruel and heartless," "absurdly wrong-headed," "antiparent," "ridiculous," and a "terrible infringement." Parents held public protest rallies where they carried signs that read "Bloomberg and Klein to NYC Schoolchildren: Don't Phone Home!" or "Mobile Phones=Security." One parent said point blank, "The chancellor will have civil disobedience on his hands. No one in New York is going to let their child go to school without a cell phone."

Even as parents push to keep phones in the hands of their children, teachers are genuinely troubled by the arrival of personal media in classrooms already struggling to keep students interested in school and learning. Talk to just about any high school teacher in schools that allow mobile phones, and their views are likely to resemble the ones that I hear repeatedly: personal media are transforming our nation's classrooms and not necessarily for the better.

■　■　■

Teachers are quite outspoken when it comes to personal media in the classroom. Among those I met, none was more outspoken than Mr. Carter, a social studies and tenth-grade pre-advanced-placement instructor in a school whose students come predominantly from working-class households.

I met with Mr. Carter on three different occasions. He was young, passionate, and in tune with his students. He cared about them and treated some as if they were his personal responsibility. Some of his students did not hesitate to text message him about an assignment, but when it came to mobile phones in the classroom, he drew a sharp line.

Walk into Mr. Carter's classroom and one of the first things you'll see is a sign that has a picture of a mobile phone with a red circle around it and a slash through it that reads, TURN IT OFF! "We [teachers] hate cell phones and iPods at the school," he told me, adding, "They are a tremendous distraction among our students. They text in class and often to one another within the school." According to Mr. Carter, his students are especially skilled at text messaging. He claims that they can send small messages via their phone without even looking at the keypad. Shaking his head in pure amazement, Mr. Carter explained, "They have gotten so sneaky and adept at texting that they can do it with the phones in their pockets." Forty-two percent of the teens in the online panel conducted by Harris Interactive said they could text blindfolded.

In addition to texting each other, teens use mobile phones to manage a wide assortment of content—ring tones, pictures, games, and videos. Phones are especially popular for accessing fast entertainment. In one conversation with tenth graders, many of them talked about using their phones while in class. Fifteen-year-old Manuel said, "Sometimes I use MySpace on my phone just to check and see what is going on." Monique, also fifteen, said, "It's hard not to at least check your phone while you are sitting in class."

The policy at Mr. Carter's school states that the use of phones in class is strictly prohibited. There is policy and then there is reality. Many students simply ignore the policy and willingly risk their phones being picked up by a teacher. Phones confiscated in this particular school district have to be retrieved by parents who are also required to pay a $15 fine. Not even the threat of a triple penalty that involves the confiscation of the phone, parents, and a fine is an effective enough deterrent to phone use in the classroom. In the first week of the 2008–09 academic year, fines for telephones exceeded $3,000 at the school.

"This may surprise you, but some parents actually call their children while they are in class," Mr. Carter said. Such behavior, he explained, makes the enforcement of the no-phone policy even more difficult. "I had a case once," Mr. Carter said, "where I got a message from the office stating that a student in my class was going home because she was sick." Mr. Carter had no idea what was happening until the student left his class and went home with her mother.

"She texted her mother and said that she was ill," Mr. Carter explained.

When the student returned to class the next day, the teacher reiterated the no-phone policy.

The ninth-grade student responded, "I'm pregnant."

"I don't care! You don't use your phone in class," Mr. Carter replied.

I knew that he cared about the young woman's welfare. Privately, he even expressed concern for her. "She's a very bright young woman, but now I'm not sure what this means for her future."

Mr. Carter also cared about his students respecting the rules regarding technology and the integrity of the schoolroom-learning environment he so desperately wanted to protect. Mobile phones in the classroom push him to the brink. So does the iPod. "They carry them around like gold," said Mr. Carter, referring to the wildly popular digital music player.

"They sit in class, and if you don't watch them, they slide the ear buds into their ears and listen to music while you are teaching or while they are supposed to be working," Mr. Carter explained to me one day. "They claim that listening to music helps them work better, which I don't always doubt, but it is another non-class-related matter that we have to deal with." Mr. Carter is not antitechnology. He exchanges MySpace and text messages with his students; some of the teachers in his school incorporate podcasts from the iPod in their instruction. Still, he believes that personal media has changed the dynamics of the classroom.

"All of this," he said, referring to repeated instances of having to deal with media in class, "is often done at the expense of stopping class and interrupting teaching and learning." Like a lot of teachers, he is frustrated by the degree to which personal media use in the classroom disrupts learning for everyone in the classroom, not just the student using media. Mr. Carter's concerns have merit.

In a 2007 study with Microsoft employees, researchers Shamsi T. Iqbal and Eric Horvitz conducted a field study to investigate the disruption and recovery of work-related tasks. Here is one of the hypotheses that they tested: employees using their computers for work will have difficulty resuming work-related tasks once they are interrupted by an e-mail or messenger service alert. To test their hypotheses, Iqbal and Horvitz studied the computers of twenty-seven employees who held a number of job descriptions, including, for example, program manager, researcher, and software developer. The researchers were able to watch the users' interaction with software applications and their associated windows in addition to a log of incoming e-mail and instant-messaging alerts. Over a two-week period, Iqbal and Horvitz collected 2,267 hours of activity data. What did all that data show?

Well, on an hourly basis a primary task—or what Iqbal and Horvitz define as the "normal daily tasks that users perform as their primary responsibility while in the computing environment"—was interrupted by an average of four e-mail alerts and three IM alerts. Many of the study participants responded within seconds to either e-mail or IM alerts, causing an interruption in the execution of their primary task. Some of the users indicated that they felt an obligation to respond to IM because someone else was on the other end. And given that e-mails are part of the workday flow, users also felt a need to respond to those alerts. According to Iqbal and Horvitz, users spent about ten minutes on the e-mail or IM switches caused by the incoming messages. But users did not stop there. Many of them tended to browse through other peripheral applications, thus further delaying the resumption of their primary task. In total, switching from a primary task to an e-mail or IM alert led to a loss of about twenty to twenty-five minutes. A main conclusion of the study is summed up this way: "Even when users respond immediately with the intention of resuming the suspended current task as soon as possible, they often end up taking significantly more time to return than the time to respond."

A number of employees repeatedly losing as many as twenty minutes of time in the workplace can be costly. In the classroom, where time and resources are already limited, the regular loss of even a few minutes can be disastrous. What seems like a negligible interruption from a student's perspective, sending and receiving small messages in class can end up costing valuable class time and the focus needed to learn a concept or complete an assignment.

There are two kinds of technologies in today's classroom: technologies that pull students away from the classroom, and technologies that pull students into the classroom. Whereas the former potentially undermine the learning experience, the latter may in turn enrich academic efforts. So far I have only discussed the former. To suggest that technology in the classroom is only a distraction would be both unfair and untrue. Across the United States there are educators who are experimenting in wonderful ways with technology to enliven the learning experience, using everything from blogs and podcasts to games and interactive maps. During a visit to an affluent seventh- through twelfth-grade school, I met Ms. Johnson, a librarian/technology coordinator who maintains constant contact with her students via blogs, her library Web site, e-mail, and IM.

When I visited her school's library Web site, I was genuinely impressed with the "integrated assignments" she helps teachers build online. On one assignment about capital punishment, the librarian used pictures of a lethal-injection facility, links to Supreme Court documents, online databases, and source aids to bring the issue and the assignment to life. Another assignment, titled "The Islam Project," used a wiki that encouraged each group to cultivate and share ideas about their projects.

"How receptive are the students to the ways you incorporate technology into their assignments?" I asked.

"The kids are amazing," she answered. "Technology is a matter of course with them. It is the way they do business."

Earlier that year she asked the ninth graders to compare a technology-based biology assignment with a more traditional assignment that included lectures, reading, and an exam. Most students chose the interactive assignment. "Ninety-five percent of the freshman class overwhelmingly and insightfully preferred it," she told me, adding that the students indicated that "they would remember the material better, as they learned it collaboratively and through a variety of sources, instead of their teacher delivering it in a lecture that they would soon forget."

Ms. Paul, another high school principal I met, is also buoyant about technology.

"When it comes to technology," she said, "the kids teach us."

The veteran administrator had recently concluded that many of her students were so skilled at making digital media content that her school had to upgrade the courses and instruction the curricula offers in media production. She also believes that technology, especially the Internet and SMART Boards, are turning the classroom into a spirited learning space. And it is not simply the technology but rather how the students and teachers are using the tools that excites Ms. Paul. "Technology in the classroom," she told me, "has reinvigorated some of my most established teachers, creating what can be called a strategic partnership between teachers and students."

Technology, Ms. Paul reminded me, is not only beneficial to students; it has payoffs for the teachers too. One thirty-year veteran teacher of government told her, "I am not retiring for several more years." For this particular teacher, discovering technology's impact in the classroom— through video, interactive maps, virtual tours and museums—was like discovering the fountain of youth. "In her class," Ms. Paul explained, "students were learning about government and campaigns in ways that were engaging, lively, relevant, and much more powerful."

During a visit to a third-grade math class, I witnessed up close how anytime, anywhere technologies can enhance the learning experiences of young children in some rather extraordinary ways.

The teacher instructed the students to gather around on the floor in front of a large SMART Board, an interactive blackboard that you can write on as well as project Web-based images and video. After working through several math problem-solving strategies, the teacher then asked the students to return to their desks, where they were given a set of instructions and problems to begin solving. Before they began working, the teacher projected a YouTube video on the SMART Board. The video was a tutorial that reiterated some of the math strategies they just covered. It was a unique way to reinforce the teacher's lesson, almost virtual and something akin to a smart game. I noticed how the SMART Board and the YouTube video kept the kids engaged and immersed in the instructional moment. In this affluent school, many of the students are accustomed to SMART Technologies' interfaces. Shortly after students began working on a few problems, class time expired and they were instructed to line up in preparation to move on to the next class. Standing in line, a third-grade girl raised her hand.

"Ms. Harris, can we watch the YouTube video you showed us at home?"

"Sure," Ms. Harris answered back. "I will send the link to your parents and they can help you pull up the video you just saw."

Ms. Harris and I looked at each other and smiled. Rather than wait until the next class day to resume working on the math problems, the eight-year-old student was savvy enough to understand that the Internet gave her another option: she could replicate part of the classroom experience at home. Sure, anytime, anywhere technologies can bring fast entertainment into the classroom in ways that are academically disruptive. But those same technologies can also bring the classroom into the home in ways that are academically productive. The key of course is figuring out ways to make the latter scenario the rule if, as of now, it is the exception in far too many instances.

The call to bring new digital media technologies—laptops, the Internet, and social Web applications—into the classroom is growing more robust. This particular movement is part of a more enduring call to modernize the nation's educational system. Driving the rationale for more experimentation in the classroom is a relatively straightforward notion: the belief that new digital media technologies can make the classroom a more relevant and stimulating place to be. In some circles the call for renovating the curricula in America's schools even includes games, a technology long regarded as a foe to learning by many educators and social scientists. Proponents of games, like literacy scholar James Paul Gee, maintain that the emphasis on interactivity, nonlinear learning, and problem solving offers good learning principles. What new digital-media technologies ultimately establish in the classroom, advocates urge, is active (doing) rather than passive (telling) learning.

Middle and high schools are not the only formal learning environments facing the unintended consequences of mobile technologies, fast entertainment, and the rampant rise of continuous partial attention (CPA). On a more personal note, I regularly witness how media multitasking in class and constantly divided attention spans are also changing the learning environment in our nation's colleges and universities. If high school teachers dread the presence of mobile phones and iPods in the classroom, a steady rising number of college professors dread the presence of laptops and classrooms offering wi-fi connections.

Before I began writing this book, one of my colleagues decided to ban laptops from his classroom. I thought the decision was heavy handed, out of touch, and unnecessary. Around the country, many college faculties are banning laptops. Now, I am learning the importance of establishing some clear rules regarding the use of laptops in the classroom. The anytime, anywhere capabilities of laptops are real and the pull of fast entertainment is relentless. CPA is just as much a part of today's classroom as teachers and students are.

One day I decided to conduct a small informal ethnographic exercise. I sat in on a large freshman introductory class of about 250 students. Sitting in the back of the large auditorium, I pulled out a notebook and began taking notes. Nobody noticed me, but it did not take long before I noticed some interesting behaviors. At least a third of the class had laptops, and toward the back of the auditorium it was clear that class time was the right time for anytime, anywhere media. Several students pulled up Facebook, and I even noticed a couple of students browsing through YouTube. Some students were using their laptops to follow class, but many others were not. Many of the students with laptops were absent in their presence.

Similar behavior happens in small classrooms too. While lecturing to my class of about thirty students one day, I noticed a student smiling as she typed on her laptop. Her delighted gaze never strayed from her computer screen, and clearly she was not smiling about my lecture on the digital divide—the gap between the technology rich and technology poor. In other words, she was in class physically but not mentally or intellectually. When I met with a small group of graduate teaching assistants, they explained that in classes held in large-size auditoriums, it is not uncommon for at least a third of the class to be using laptops. It is impossible to know what students are doing on their computers in large- and smaller-size classes.

Continuous partial attention in the classroom not only raises questions about what, if anything, students are learning when they divide their attention between class and fast entertainment. It also raises questions about the overall learning.environment in the college classroom today. Back in the day, students' attention certainly wandered in the classroom; they daydreamed, doodled in their notebooks, read a magazine, or simply fell asleep. And there were certainly distractions in the lecture hall before laptops and the Web, including newspapers, magazines, crossword puzzles, and portable music players, just to name a few. There is growing suspicion, however, that mobile technologies enhances the degree of intensity with which students' attention wanders today. Imagine a classroom in which one-third of the students pull out newspapers to read. In many ways, checking e-mail, browsing Web sites, or checking for status updates on Facebook while in a classroom is the digital-world equivalent of that scene. Apparently, the lack of focus has been particularly acute in many law school classrooms.

Citing their disruptive nature, a number of law professors have banned the use of laptops in their classes. A growing number of legal scholars are finding it difficult to practice the Socratic method, a widely used teaching technique in legal education, in classrooms offering wi-fi. The Socratic method works best when students are engaged, poised, and ready to discuss legal cases, theories, and history with focused intensity. In a *Washington Post* op-ed piece, David Cole, a Georgetown University law professor, writes that the response he hears most often from his students when practicing the Socratic method is, "Can you repeat the question, please?" Cole adds, "It is usually asked while the student glances up from the laptop screen that otherwise occupies his or her field of vision."

When June Entman banned laptops in her classes at the University of Memphis Law School, students staged a miniprotest by passing around a petition to regain the right to use their laptops. In an e-mail to her students, Entman explained the ban this way. "The wall of vertical screens keeps me from seeing many of your faces, even those of some students who are only neighbors of a laptop." Entman maintains that laptop screens hamper the flow of discussion and thus diminish the quality of the learning experience.

Another personal encounter I had in the classroom is also instructive. After noticing how a graduate student's use of his laptop was causing him to drift in and out of the seminar, I sent an e-mail to the class reminding them that if they use a computer during class it should be related to their involvement in the class. My e-mail sparked an interesting discussion the next time the class met. A couple of students in this small seminar politely made the case that using a laptop or some other device in class to check e-mail or attend to some other non-class-related task did not weaken the quality of the learning environment or their performance in the class. The student whose behavior provoked my initial e-mail even went so far as to say that it was impossible to sit through a near three-hour class (we do take breaks) without his laptop and presumably the ability to browse the Web. The conversation was a revelation.

Asking young people to disconnect even momentarily from the vast swirl of content and comrades they engage throughout the day generates anxiety, discomfort, and cultural alienation. Today, when you ask students to turn off their computers, mobile phones, and iPods, you are asking them to turn off their lives. The pushback from the Memphis law students or the two graduate students in my class is in many ways a bid to stay alive. Nothing more, nothing less.

After pondering about the exchange with the students, I asked myself, Is it even reasonable to expect students to commit their full attention in a classroom setting? Have we created a culture in which the ability to pay attention to a single thing for a sustained period of time is simply no longer possible in a age of constant stimulation, communication, and gratification? Anyone managing a classroom today can no longer ignore the reality of continuous partial attention and the pervasiveness of personal media. Some observers believe that the hyperkinetic, always-on environment in which we live makes it next to impossible to maintain focus for extended periods of time. Dr. Edward Hallowell, a psychiatrist who specializes in attention deficit disorder (ADD), subscribes to this view.

Hallowell, a former Harvard Medical School professor, describes this condition as attention deficit trait (ADT). He asserts that while humans are born with ADD, ADT is environmental. Discussing ADT, Hallowell says, "It's a condition induced by modern life, in which you've become so busy attending to so many inputs and outputs that you become increasingly distracted, irritable, impulsive, restless, and, over the long term, underachieving." No matter if it's adults in the corporate world or students in the classroom, ADT, Hallowell contends, leads to brain labor overload and a dilution of our mental powers and performance capabilities.

To be fair, the cultural changes partially wrought by technological changes demand that instructors rethink their approach to pedagogy and learning. Technology thought leader Don Tapscott likens the traditional model of teaching—the lecture—to the broadcast. A broadcast, he reminds us, "is by definition the transmission of information from transmitter to receiver in a one-way, linear fashion." Both television and the lecture, Tapscott argues, are losing their audience. "Sitting in front of a TV set—or a teacher—doesn't appeal to this generation," he writes. "But unlike the entertainment world, the educational establishment doesn't offer enough alternatives to the one-way broadcast."

Maybe students are surfing the Web in class because they are not being sufficiently engaged in class. Or, maybe they are surfing the Web in class because CPA makes it increasingly difficult to sustain their focus. Either way, the culture is evolving and so are learners. How instructors approach students and learning must evolve too. Technology in the end is never the problem or the solution. Humans are.

If you think the concerns about multitasking, fast entertainment, and CPA are simply a generational gap between the so-called digital natives and the digital immigrants, think again. Talk to young people as we did and you'll discover that many of them are also grappling with the consequences of anytime, anywhere technology.

On practically every scale that we used to examine young people's attitudes about social and mobile media, the outcome is unequivocally favorable. They use, trust, and derive immense satisfaction from their laptops, mobile phones, and the Web. More than half of those we surveyed, 56 percent, believe the Web is a necessity in life. And it is true. Every facet of their lives—school, work, play, finances, shopping, and communication with their close friends—is managed through the Web.

And yet, when we ask young twenty-somethings if there are any regrettable aspects related to their engagement with social and mobile media, they consistently point to one factor: the difficulty in turning it off when they need to focus on school or work. Despite college students' fascination, for instance, with Facebook, they routinely described the platform as "a major time-suck." Melanie, a twenty-one-year-old public relations major, said that Facebook distracts her from her homework. Twenty-year-old Scott believes that he would do more worthwhile things if he did not spend so much time on the popular platform. Josh spends a lot of his day chatting with friends online and checking their Facebook status.

"I spend way too much time [online]," Josh said, "and I believe they [social-network sites] are addicting because you can get on Facebook and not realize how much time has passed. You look up and it can be hours later."

Echoing this concern about the Web, twenty-two-year old Byron told us, "You can go on there and lose three hours of your time when you should be studying instead of viewing other people's profiles." The ability to see what their friends are doing, along with whom they are doing it with, is not only a powerful pull, it is a constant pull. It is not that social and mobile media are necessarily addictive but rather that they are omnipresent.

For many students the ability to maintain their focus and attention on an academic task is especially challenging when the tool they are likely using to execute the task—a computer—is their screen of choice for information, entertainment, and life sharing with their friends. In fact, many of the college students we spoke with readily acknowledge that the Internet is a major source of distraction while trying to study. This is not a recent trend. In a 1997 survey of 531 college students, psychology professor Kathy Scherer reports that 13 percent of her respondents acknowledged that excessive use of the Internet interfered in their social life and/or academic performance. While it is not recent, the impact of the Web on academic performance is evolving and in some ways intensifying. The constant connectivity afforded by today's technology has become a highly desirable feature of our digital environment. In our survey, more than four in ten, or 44 percent, agreed with the statement "I cannot stay away from the Internet for too long." What they really can't seem to stay away from are the opportunities to communicate with their friends or check their profiles for updates. I gained a newfound appreciation for this small factoid after reading a student's paper one day.

This particular student was writing about her peer's use of social-network sites. As I read the paper, I noticed several unexplained asterisks. Unsure what the marks meant, I continued on. Finally, at the end of her ten-page report, the student explained the asterisks: they signaled each time she logged on to Facebook while writing her paper. She stopped six times in the middle of the paper to check Facebook. Interestingly, the class was examining the social consequences of the social Web, and that prompted her to write this about her favorite online site: "I think that it is a very good way to show not only how entertaining it can be, but how distracting it can be!"

Jennifer, a twenty-six-year-old sociology major, is like most young people that we met. The Web is a part of her day, "pretty much throughout the day." She prefers MySpace. "I usually use it as a break from studying. Check my e-mail or check the news," she said. Usually when she pulls up MySpace for a quick study break she finds herself staying online much longer than she intended. "I can't help myself," she explained. Scanning the profiles of friends and acquaintances for simple amusement has become so habitual that many students find themselves constantly battling the urge to do so even when they are sitting in class or trying to complete school-related tasks. When Jennifer absolutely needs to get some schoolwork done, she physically separates herself from the computer. "I sometimes try to leave my computer home on purpose and go to coffee shops so I won't use it because it's such a draw," she said. As extreme as this may sound, it is not out of the ordinary.

We heard an assortment of ways young people fight back the constant pull of online videos, games, Facebook, and all of the other bite-size digital goodies when they need to study. One of the most urgent matters in today's always-on lifestyle is not addiction but rather the division of our attention.

Twenty-one-year-old Jake, a history major, also finds it difficult to maintain his focus on schoolwork when he is near his computer. "I find myself having to physically remove myself from the computer in order to read or write a paper," he admitted.

Jake continued: "Like, my computer stays in my living room so my roommate and I can use it, and I'll have to go into my room and close the door to do reading. Like, physically get up away from the computer and, like, sit on the couch so I'm not next to the computer." Otherwise, he claimed, "I'll just pass time on the computer."

Not that long ago, television used to be our chief source for passing time. Not anymore. Whereas 27 percent of those we surveyed said they use television "most often" to pass time, twice as many, 54 percent, said they use the Internet to pass time. A Pew study reports that 82 percent of persons ages eighteen to twenty-nine have gone online just to pass time. When asked if they did it yesterday, 37 percent said yes.

Elizabeth also uses creative methods to resist her computer and the allure of fast entertainment. The twenty-two-year-old government and fine arts major said, "It takes a lot of willpower for me to not check it [Facebook] a lot when I'm doing my homework because homework is a lot more boring. . . . It's the truth."

We have every reason to believe that young people like these are increasingly the norm rather than the exception. It is clear that they are looking for a place of refuge from the deluge of content accessible through their computers. They are also looking for a place to simply stop, disconnect, and think. Hallowell maintains that "if you don't allow yourself to stop and think, you're not getting the best of your brain. What your brain is best equipped to do is to think, to analyze, to dissect and create. And if you're simply responding to bits of stimulation, you won't ever go deep." Many students, I believe, would agree.

While we were out in the field talking with young people, Facebook launched its news-feed feature that automatically alerts users of any and all changes made by someone in their network. While some Facebook users complained that the feature was too invasive, many others eagerly embraced it. Elizabeth absolutely loves the news feed. She admitted, "And even if nothing's changed, I think it's almost even just the habit of thinking that something exciting might be happening on the Internet. That's more exciting than the paper that I'm writing."

Several of the young people that we talked to acknowledged that they are constantly logging on to see if anyone has posted comments on their wall, uploaded new pictures, or changed their status update. This constant state of alert, the sense that in an always-on world you are missing something, is, of course, one of the key symptoms of continuous partial attention.

More than half of the young people we surveyed said that they check Facebook three or more times a day. In our in-depth conversations it is clear that many students check several times throughout the day. Kayla, a twenty-two-year-old economics major, could not count. Her response to the question neatly summed up what we often heard: "I check my Facebook at least once every hour at work," she said. "Something is always happening on Facebook."

As important as the debate about addiction, multitasking, and CPA is, something subtle yet equally profound is occurring alongside the widespread diffusion of social and mobile media. Anytime, anywhere technologies do more than deliver fast entertainment and steal away our attention. These technologies do more than transform how we consume content. Social- and mobile-media platforms also transform our world by transforming how we experience space. In an environment where fast entertainment is always accessible, the boundaries between traditional leisure spaces (think home or the cinema) and nonleisure places (think work or school) are erased. In today's technology-rich world, any place can be a leisure space—a place to download a video, watch a movie clip, listen to your favorite pop single, or take a quick peek at a friend's personal profile. Addiction is not the more common challenge in the digital world; the ubiquitous presence of entertainment and the desire for constant gratification are.

In the current cultural milieu, fast entertainment is more than a luxury or a way to pass time. It is an entitlement that more and more of us expect no matter where we are—at home, at work, in school, on vacation, or even when driving our cars. That cultural ethos, or the expectation that anytime is the right time for entertainment, is transforming our behavior and our world.

A Message from Barack

What the Young and the Digital
Means for Our Political Future

Young people are on the Web. That's how we're organizing.
—Meredith Segal, founder of Students for Barack Obama

On February 2, 2007, Illinois Senator Barack Obama spoke at the Democratic National Committee winter meeting held at the Hilton Washington Hotel. Over the next two days, each of the hopefuls for the Democratic presidential nomination spoke to the body of elected officials and party enthusiasts. After finishing his commitments with the party establishment, Obama and a few of his staff hopped into a car and rode over to a more raucous gathering a few miles away in Fairfax, Virginia. A group called Students for Barack Obama had invited the senator to speak at an afternoon rally they were holding in support of his impending candidacy at George Mason University. It would be eight more days before Obama officially announced his bid for the Democratic Party nomination, but that did not deter young America. Almost two years before the election, they were fired up about his potential historic run.

On the way over to the event, Obama turned to one of his staff members and asked, "About how many people do we expect to be there?" Unsure, the staffer responded, "Oh, about a hundred or so." The staffers thought Obama could make an appearance, offer some words of encouragement, and thank the students for their early support. The visit to George Mason turned out to be much more than that.

When Obama and his staff walked into the Johnson Center Atrium, they were stunned by what they saw. The venue was packed wall to wall with bodies that scaled all the way up the spiraling, five-story atrium. The building could barely contain the energy generated by the estimated 3,500 people in attendance. Many of the students had been camping out since the early morning hours to assure themselves a good spot at the afternoon event. They waved signs, cheered wildly, and snapped pictures of the man they wanted as their next president. Looking at the crowd with genuine amazement and gratitude in his eyes, Obama told them, "This is a remarkable achievement, a remarkable event that speaks to what's possible when young people put their minds to something."

On that cold winter day in Fairfax, Obama learned an early and—as it turns out—pivotal lesson about the power of the young and the digital. It was, in retrospect, his very own digital awakening.

Appearing at a campaign rally in Los Angeles a few weeks later, the senator spoke fondly of the George Mason event. "About three weeks back there were a group of students who had decided to start an organization called 'Students for Barack Obama.' And they used this thing called Facebook," he told an outdoor crowd. Prior to the George Mason event, Obama did not know much about Facebook or social media. He admitted as much during the campaign event in Los Angeles. "I don't really understand it," Obama said. "Facebook is a young person's thing." With the George Mason event still fresh in his mind, Obama referred to it as "an astonishing site and they had done it all on their own. And without us [his staff] doing anything."

A great majority of the students found out about the event through the online social networks that are a common part of their daily lives. Jonathan Hicks, a nineteen-year-old sophomore attending American University, found out through Facebook. "Technology is changing. Politicians need to use it more, and more often, if they want to reach the youth of America today," Hicks told a *Washington Post* reporter covering the event.

Two of the students that Obama thanked during the George Mason event, Meredith Segal and Farouk Olu Aregbe, helped build the online infrastructure of college students supporting his candidacy. In July 2006, Segal, a junior at Bowdoin College at the time, created the Facebook group Students for Barack Obama. When Obama visited George Mason eight months later, eighty chapters of Students for Barack Obama, consisting of more than 62,000 students, had been created. It was a powerful army of young college activists. Aregbe was one of those who joined Students for Barack Obama. But he did not stop there. He went on to create a group within Facebook called One Million Strong for Barack. In the first hour the group had one hundred members. In less than five days the group's numbers rose to ten thousand. By the third week the group's numbers approached two hundred thousand. Aregbe's was one of five hundred "groups" supporting Obama that were created on Facebook.

By the time Obama was sworn in as the forty-fourth president of the United States, more than 940,000 members had joined One Million Strong for Barack. Facebook became a place to express their thoughts about the candidates *and* participate in the process. Over 64,000 discussion topics were circulated through the group. In the true spirit of the young and the digital, they posted over 4,000 photos as well as an assortment of videos and links to new stories, events, and other related matters.

When Web evangelists talk about the power of the networked world, it is in reference to statistics like those above. Networks have always been key sources of power, providing access to social and financial capital. In the digital age, our social networks—how they form and what they do—are changing. First, our networks simply multiply faster and scale larger than anything humans have ever experienced. Not only are our networks bigger, but as Don Tapscott explains, they are also more complex and more efficient. Tapscott refers to these connections as "N-Fluence networks," arguing that Net Geners use them to "discuss brands, companies, products, and services." Young Obama supporters certainly used their online social networks to discuss their candidate. The social networks enhanced by social and mobile media have made the online communities built by users of digital technologies a vital source of power, community, and influence.

Obama is not only smart; he is also a quick learner. All of the candidates knew that the Web was a great fundraising tool. Howard Dean proved you could raise millions of dollars through hundreds of thousands of small donations via the Internet. In that crowd at George Mason University, however, Obama saw something that none of the other presidential contenders saw: the power of digital as a grassroots tool and a social tool. He saw the future of electoral politics.

Once he realized the looming impact of the young and the digital, Obama did two things. First, he hired some of the top young technology talent to design and execute his new media strategy. Second, he established an active presence in the online world.

At the heart of Obama's new media team was a core of young technology experts who understood that for the thirty and under set, digital is more than another tool or tactic, it is a way of life. Obama enlisted the support of young technology thought leaders like Joe Rospars, who, in his early twenties, was one of the central players in Howard Dean's inventive use of the Web in 2004. Rospars went on to become a founding member of Blue State Digital, widely regarded as a cutting-edge new media political consultancy.

Team Obama also recruited Chris Hughes. No one understood the power of online social-networking tools more than Hughes. Though only twenty-four when he joined Obama's campaign, Hughes had accumulated crucial experience by watching up close the explosive growth of online social networks. As a cofounder of Facebook, he played a key role in building an online platform that was changing how young people maintained their social ties and managed their lives.

Hughes used his considerable talents to help the candidate build his own social-network platform, MyBarackObama. The site was a brilliant example of the power of social media. By the end of the campaign, 1.4 million users had set up accounts and helped to form a digital army of supporters that was unprecedented in its scope and campaign involvement. Brian Stelter writes that Hughes "wanted Mr. Obama's social network to mirror the off-line world the same way that Facebook seeks to, because supporters would foster more meaningful connections by attending neighborhood meetings and calling on people who were part of their daily lives." Sociologist Barry Wellman calls the social connections people build and maintain through the Web "personal communities." The more personal the online connections are, the more meaningful they are. And the more meaningful online connections are, the more likely people will actually engage them.

The young squad that managed Obama's social-network site affectionately referred to it as MyBO. Team Obama did not use MyBO to simply post campaign content or ask visitors to make a donation. All of the candidates did that. MyBO went further by asking users to create and share their own stories, build and engage with their own networks, and organize their own local campaign efforts and events. The true spirit of the platform, users connecting locally with family, friends, and neighbors, resonates with how young people use social media everyday to connect with their peers.

By turning his new media operations over to a group of young tech creatives and designers, Obama also turned a corner in American electoral politics. Obama's rise, on many dimensions, represents a generational passing of the torch. This is vividly evident in the way that he incorporated technology into his campaign. Joe Trippi, Dean's campaign manager in 2004, and no stranger to how new technologies impact electoral strategy, offers this assessment: "Just like Kennedy brought in the television presidency, I think we're about to see the first wired, connected, networked presidency."

After hiring top talent from the online world, Obama did something else—he began participating in the online world himself. Obama did not create an online movement, he joined one. He did not know much about social media at the start of his improbable journey, but by the end Obama was everywhere in the digital universe. In addition to his own social-network site, Obama established a presence in places like MySpace, Facebook, YouTube, Twitter, Black Planet, MiGente, LinkedIn, Digg, and Flickr, the online photo-sharing community. To reach young men, Obama did something no other presidential candidate had ever done—he went into video games. Whereas most politicians criticize games, Team Obama purchased real-time ads encouraging early voting in popular titles like *Guitar Hero III*, *Nascar 09*, *NBA Live 09*, and *Madden NFL 09*.

Appearing at the Web 2.0 Summit in San Francisco shortly after Obama's victory, Arianna Huffington declared that, "were it not for the Internet, Barack Obama would not be president. Were it not for the Internet, Barack Obama would not have been the nominee." The editor in chief of the *Huffington Post* was half right. Yes, new media technologies certainly aided Obama's historic run. But he was not successful because of the Internet. He was successful because of how he incorporated the Internet into his campaign. It was no great secret that the Web would play a role in the 2008 presidential election. And yet Obama's new media team understood better than anyone else's that it was not simply the technology that was important but rather understanding and then leveraging how young people use technology in their daily lives.

From the beginning Team Obama understood that digital was not merely a tool to target young citizens but rather a medium to talk with them, open up to them, and interact with them. By enlisting young technology users to lead his new media strategy, Obama came to understand the things we are learning through our research with young technology users. Technology, first and foremost, is social and communal in their world. Further, young people use social and mobile media to manage nearly every aspect of their lives—being digital is simply the way they live. Importantly, teens and young twenty-somethings use technology to share their lives with each other.

On the one hand, Obama's campaign was unquestionably traditional —tightly controlled, formal, corporate, capitalist, and top down. His use of new media, on the other hand, meant that an element of the campaign was exquisitely nontraditional—open, intimate, interactive, casual, and bottom up. Plunging fully into social media certainly poses risks for candidates. It exposes them to the flaming, cyberbullying, and hate speech that frequently occur in the digital public sphere. Further, it means that they don't always control the representation of their image or message—technology users do. But diving into the social-media pool has its rewards too. Social media allows candidates to sustain a conversation with supporters. Equally important, it allows supporters to talk with each other about the candidates. Blogger Ranjit Mathoda notes that the Web also allows campaigns "to lower the cost of building a political brand, create a sense of connection and engagement, and dispense with the command and control method of governing to allow people to self-organize to do the work."

In February 2007 Obama met with Marc Andreessen, a founder of

Netscape and a Silicon Valley insider. Obama was interested in discussing how new technologies, especially online social networks, could aid his presidential bid. Obama, Andreessen thought, was like a technology start-up: small but smart, inexperienced but entrepreneurial. "It was like a guy in a garage who was thinking of taking on the biggest names in the business," Andreessen told a reporter. "What he was doing shouldn't have been possible, but we see a lot of that out here and then something clicks."

Andreessen came away from the meeting convinced that Obama was a different kind of candidate. "Other politicians I have met with are always impressed by the Web and surprised by what it could do, but their interest sort of ended in how much money you could raise." Obama, Andreessen concluded, "was the first politician I dealt with who understood that the technology was a given and that it could be used in new ways."

It was the use of the Web in those news ways, primarily social and communal, that distinguished Obama's new media strategy from his opponents. All of the candidates used new media, but Obama used it the way young people did—casually and socially. He used it to share parts of himself and his campaign with voters. Through e-mails, text messages, Twitter updates, YouTube videos, and other platforms, Obama maintained a constant connection to young voters.

Even after winning the election, Obama used new media to stay connected to voters.

Minutes before appearing in front of an ecstatic Chicago crowd ready to celebrate his incredible victory, Obama sent a message to his massive e-mail list. "I'm about to head to Grant Park to talk to everyone gathered there, but I wanted to write to you first," the message began. He recognized his supporters: "You made history every single day during this campaign—every day you knocked on doors, made a donation, or talked to your family, friends, and neighbors about why you believe it's time for change." Obama also vowed to keep the conversation going. "We have a lot of work to do to get our country back on track, and I'll be in touch soon about what comes next."

The e-mail was typical Obama—well timed, direct, and personal. Throughout the campaign Obama and top members of his team regularly sent e-mails and text messages to the millions of people who joined his online database. Many of the e-mails from the candidate contained the simple subject heading "A message from Barack." The campaign's use of social and mobile media displayed all of the hallmarks of a con-stantly connected generation. Team Obama used the technology to maintain ongoing interaction with voters. This is what made Obama a Mac (cool, inspiring, and in touch with the times) and his opponents a PC (dull, uninspiring, and out of touch).

Two days after winning the election, Team Obama released a set of intimate photos from election night on Flickr. By sharing them with the public, the Obama campaign once again displayed a willing desire to embrace the life-sharing inclinations of the young and the digital. The gesture struck a chord in the Flickr community.

"Thanks for sharing . . . all these poignant, personal glimpses into you and your family's journey," one Flickr user said. Another member of the photo-sharing community referred to the gesture simply as "amaz-ing openness."

To many of Flickr's users, the pictures were familiar and foreign at the same time. Sharing intimate moments in such a quasi-public man-ner is certainly familiar to them. The fact that the newly elected presi-dent of the United States would do so, however, was absolutely foreign. Life sharing by the president-elect pointed to a new brand of politics, indeed, a new political future. One Flickr user sums it up best: "Sorry—but I wouldn't have thought to see the elected President of the US on flickr in such a private pose on such a private-public event. This is 2008, obviously . . . Mr. Obama seems so normal-natural-human. Let's hope he proves true of a millions' dreams."

Team Obama's approach to new media—social, casual, and personal —generated real credibility among young voters. Peter Hart, a widely respected and longtime pollster of the American electorate, marveled at Obama's appeal to the thirty and under set. "Young voters," Hart explains, "were drawn to him from day one and it was a connection that was as psychological as it was issue driven."

YouTube is a favorite online destination among the young and the digi-tal. The platform even emerged as their unofficial network for watching and sharing campaign-related news, events, and media. Obama's new media team used viral media expertly, uploading more than 1,800 YouTube videos over the course of his presidential run. That's a lot of content. By several accounts, considerable portions of it were viewed. In the final two weeks of the campaign, BarackObamaTV was one of the most viewed channels on YouTube. Statistics from TubeMogul.com show that on election eve, November 3, 2008, BarackObama.com generated 1,477, 579 views compared to 163,940 views for his Republican opponent, John McCain. Obama's YouTube views—like his money, television ads, and army of ground troops—dwarfed McCain.

Micah Sifry, from techPresident.com, asked TubeMogul to calculate how much time viewers spent watching the videos posted by Obama and McCain. The gap in viewing time between the two campaigns is staggering. TubeMogul estimated that the content Obama posted generated 14,548,809 hours of viewing compared to 488,093 hours for McCain. Next, Sifry asked Trippi to estimate how much it would cost to purchase this much time on television. Trippi ran some numbers using the Denver television market, a key locale in a crucial battleground state, as a model. According to Trippi, Obama would have to spend about $46 million to attract those same eyeballs on television. McCain's YouTube views translate into about $1.5 million. Noting the key difference between YouTube content and television ads, Trippi explains that "the finer point would be that people were not forced to watch these—they wanted to watch them—they chose to watch them." Sifry ends his blog piece with a simple but significant question: "Can any candidate afford to ignore YouTube in the future?"

Many journalists call 2008 the "first YouTube election." According to the *Nation*'s Ari Melber, YouTube users viewed more videos from citizens and independent groups than the two campaigns. One video in particular signaled the rise and currency of user-generated media on YouTube. Early in the primary season, pop music maestro Will.i.Am uploaded a stunning mash-up of a speech Obama delivered after losing the New Hampshire primary to Hillary Clinton. Will.i.Am turned Obama's riveting speech into a memorable music video set to simple chords, stirring lyrics, and cameos from a diverse group of pop-culture celebrities. "Yes We Can—Barack Obama Music Video" is a brilliant example of how user-generated content became its own source of citizen media and political engagement. At the time this book went to press, the video had been viewed more than 15 million times on YouTube, and according to the site it is the second most viewed video in its news and politics category.

Most of the buzz about YouTube focuses on the videos that users watch and share with each other. But YouTube is interesting for other reasons too, namely, the communities and conversations people form. The world's largest online video-sharing platform has evolved into a virtual watercooler, a place where we gather to share our thoughts about the things that we care about—politics, pop culture, religion, sports, and sex. The conversations on YouTube demonstrate quite vividly and sometimes crudely how the social Web has become a town square for people to exchange their views and opinions. *New York Times* television critic Virginia Heffernan calls YouTube "a sort of metropolitan idea that you'll run into and discover a friction with people who are very unlike you." Will.i.Am's video "Yes We Can" generated more than 87,000 comments, making it one of the most discussed videos of all time on YouTube. While the comments on YouTube are typically short, they are occasionally quite substantive. That is what Heffernan discovered in her analysis of a debate about Islam that emerged on YouTube in response to a video titled "The Truth about Islam from an Ex-Muslim Lady."

In her assessment of the community and commentary ignited by the video, Heffernan writes that "the expansive commentary is fast becoming a full-blown novel of world religion, one that dramatizes the fascinating and often shocking preoccupations of today's desk-chair ideologues."

YouTube users are not passive. Instead, they often engage the content and the commentary posted on the platform. Over the course of the campaign, YouTube became a place for voters to ask debate questions, watch the candidates, and express their thoughts about the campaign. Future presidential campaigns may not be able to control the citizen-led conversations that happen on the social Web, but one thing is certain: they must participate in them.

■ ■ ■

Robert Putnam maintains that starting around the 1970s, throughout the eighties, and into the nineties, a rising number of Americans "turned off" and "tuned out" politics. Looking at trends in presidential voting, Putnam identifies a precipitous decline between 1960 and 1996. In 1960, 62.8 percent of eligible voters went to the polls. By contrast, only 48.9 percent cast votes in 1996. Assessing the voting data, Putnam concludes that "participation in presidential elections has declined by roughly a quarter over the last thirty-six years."

Putnam also looks at other expressions of civic engagement, including signing petitions, making speeches, or writing an elected official or letter to the editor. Putnam notes, "In 1973 most Americans engaged in at least one of these forms of civic involvement every year." Twenty years later, American's civic involvement was declining sharply. "By 1994," Putnam contends, *most did not engage in any.*" Even more distressing, Putnam explains, is the decrease in forms of civic involvement that bring people together, build community, and encourage participatory democracy. These more collaborative activities include things like working for a political party, serving as an officer of an organization, or attending a political rally. "Those activities that brought citizens together, those activities that most clearly embody social capital," Putnam says, "have declined most rapidly."

But times are changing. Both the anecdotal and empirical evidence from 2008 suggests that voting, public expressions of engagement, and communal involvement in politics may be on the rebound. Along with the rise of the Web has been the ascent of new expressions of political engagement that are poised to reverse the retreat from electoral politics and civic involvement documented by Putnam. And the young and the digital are leading the way.

Americans under the age of thirty did more than sign up for Facebook groups like Students for Obama or One Million Strong for Barack. With platforms like Facebook, Twitter, and YouTube, as well as mobile phones, they could tap their networks anytime and, importantly, on their own time. They used their online networks to build an off-line grassroots movement that rallied formidable support for their candidate and created one of the richest assets a campaign can have: a citizen-led network that took an ownership stake in Obama's campaign. When millions of citizens make such a personal investment in a campaign, they become more than consumers of political content; they become creators and distributors of political content too.

Throughout the campaign young Americans displayed new expressions of citizenship. They signed online petitions and joined online organizations. Young citizens expressed their views through blogs, virtual paraphernalia, and mash-up videos. Social-network sites evolved into political communities and were used to share pictures, news links, campaign announcements, and video clips that were serious and hilarious. Sure, young people watched the endless flow of video parodies that appeared online. Tina Fey's *Saturday Night Live* skits became "Must-Stream TV." But Obama's thirty-seven-minute speech on race in response to the Jeremiah Wright controversy was one of the most shared campaign-related videos on Facebook. More than 5 million people watched the speech on YouTube. Young people also used new media to invite their peers to pass out campaign literature, set up voter registration tables, attend local campaign events and debate watch parties, and go to the polls. Consistent with what I report in the chapters above, online political activity was connected to off-line political activity. It was young people's very own version of a political ground war.

After the election the Pew Research Center asked voters if they attended campaign events. Pew's findings may surprise you: more young voters than older voters reported attending a campaign event. Throughout the course of one of the longest presidential campaigns in American history, young people defied the widespread notion that their enthusiastic embrace of technology is really a form of disengagement from the world around them. In fact the opposite was true: their embrace of technology encourages greater involvement in politics and the world around them. Here is another interesting figure. More than half, 51 percent, of his online supporters expect some form of communication from the Obama administration through e-mail, text messaging, or social-network sites, according to the Pew Research Center. That kind of interaction and networked governing represents a new generation of political participation.

In the chapters above I address what young people's engagement with new communication technologies means for how they bond, communicate, and manage their lives. The implications for how they participate in America's great democracy experiment are equally significant. Take, for instance, how young people stay informed about politics.

Television, for at least three decades, has long been the dominant source of news and information in America. In 1987 the Roper Organization reported that for the first time ever a majority of Americans, 50 percent, cited only television as their main source of news. According to the report, poll respondents believed that compared to newspapers, magazines, and radio, television was more trustworthy. Twenty years later, social, technological, and generational change was the catalyst for another transformation.

In a December 2008 poll conducted by the Pew Research Center, 40 percent of respondents said that they get most of their news from the Internet. That was a dramatic increase from the 24 percent who reported getting their news from the Internet in September 2007. For the first time ever, in a Pew survey, more people reported getting their news and information from the Internet rather than newspapers. Among all Americans television was still the primary news source, but among voters under thirty the Internet and television were about equal. Pew writes: "Nearly six-in-ten Americans younger than 30 (59%) say they get most of their national and international news online; an identical percentage cites television." In our survey of college-age persons, 62 percent said that they get their news from the Internet, whereas 28 percent said television.

Though younger Americans do not read newspapers or watch the evening news, they are usually just a click, scroll, or link away from what is happening in the world. In fact they get their news the way they get their media—from the Web, on the go, and peer to peer. In a world of anytime, anywhere media and constant connectivity, news often comes to us. One young student summed it up best: "If the news is that important, it will find me." Whatever the problems posed by social and mobile media, a lack of news and information is not one of them.

■ ■ ■

While young people's political engagement via new media in 2008 was significant, one questioned still lingered: would they turnout to vote? Many of the early indicators suggests that they did, and in noteworthy numbers. A larger percentage of Americans under the age of thirty voted than at anytime since 1972, when 55.4 percent cast a vote.

Tufts University's Center for Information and Research on Civic Learning and Engagement, or CIRCLE, estimates that youth voter turnout was between 52 and 53 percent in 2008. That represents about a 4 to 5 percent increase from 2004 estimates. In 2004, 19.6 million Americans under age thirty voted in the presidential election. Four years later, 3.4 million more young people, or 23 million, voted. The growing number of young people casting a vote in a general election is a reversal of a near quarter-of-a-century trend that produced consistently weak turnout.

A combination of factors explains the recent increases in youth voter turnout: a government that seems more detached and ineffective than ever before, an unpopular war that has claimed thousands of young American lives. Add global warming, the spiraling cost of a college degree, and the worst economic crisis since the Great Depression to the list, and the urgency of the moment is apparent. From the view of young people, Obama's global outlook, call for greater transparency in government, and quest for change made him a clear and easy choice in such gloomy and difficult times. Obama's use of social and mobile media expressed a willingness to close the gap between politicians and young America.

"One thing that's been incredibly clear throughout this whole process is his commitment and dedication to students and all the young people of America," said a college junior. "He sees our generation as a critical part of his campaign."

Obama won the support of young voters by an extraordinary margin. In 2000 Al Gore won the youth vote by two points. John Kerry fared a bit better four years later, winning that demographic by seven points. In 2008 Obama won young voters by thirty-four points. The future impact of that is beyond profound. Today's young voters represent tomorrow's electorate.

Andy Oram, an editor at O'Reilly Media, adopts a more cautious position when it comes to the impact of the Web in the 2008 presidential race. Oram claims that "elections have not been fundamentally changed by the Internet," adding that "2008 is still the era of mainstream media." Significant portions of the money Obama raised via the Web, Oram argues, were necessary to pay for what is still the most important and expensive aspect of presidential elections, television ads. The 2008 race for the White House, Oram contends, was about legacy media. Looking at the world as it is today, Oram makes a reasonable argument. Looking ahead as the world will be tomorrow, his argument loses strength.

Americans under the age of thirty made up 18 percent of the 2008 vote. If the demographic patterns from the 2008 vote stay the same in twelve years (and they will not: the general population is trending more racially and ethnically diverse and that population is significantly younger), today's young voters along with tomorrow's younger voters will make up at least 50 percent of the voting population.

Besides, in a few short years at least half of all likely voters will seldom if ever use the legacy media referred to by Oram. Their primary media will be social, mobile, and personal. Any political ad they see will likely be on a personal computer or mobile device. Any poll they participate in will not take place over a landline. The information they access about candidates is just as likely to be distributed by a peer as it is by a professional journalist. And to the degree that the behaviors of young technology users influence the older set, changes like these will impact all voters.

The migration to digital highlights a steady and defining generational shift in American life. Sociologists use the term *intercohort* to identify social, attitudinal, or behavioral changes that occur across specific age groups. As Putnam notes, "If different generations have different tastes or habits, the social physiology of birth and death will eventually transform society, *even if no individual ever changes*." Meanwhile, 2008 offers a glimpse of our political future, a soon-to-come world where campaigns and expressions of political engagement will be defined by the behaviors and sensibilities of the young and the digital. In that world, citizens will use new technologies to participate in politics, create and share their own political content, and talk to peers and candidates alike.

As his campaign took him across the country, Barack Obama grew fond of telling his supporters that "something is happening in America." And he was right. Something was definitely happening among the young people who crossed racial, ethnic, and even party lines to back him by remarkable margins. Every day in their technology-rich lives, young people embodied the mantra that drove Obama's historic candidacy—CHANGE. He saw change in the way that young citizens engaged his campaign, each other, and the world around them. Change is especially evident in the ways that the young and the digital use social and mobile media to write their own unique American story about life, liberty, and the pursuit of happiness.

THE MAKING OF THIS BOOK
RESEARCH, METHODS, AND ACKNOWLEDGMENTS

In September of 2006 I shared an interesting conversation with a researcher from NBC Universal. Like the other major television networks, NBC was on edge as the new fall season began. The audience for the major networks had been declining for years, but the level of industry anxiety was growing more palpable. This was certainly true at NBC. In a short three-year span, the network known for top-rated shows like *Seinfeld* and *Friends* had fallen from first to last place among its "big four" comrades—CBS, ABC, and FOX. Most significant at NBC was the erosion of the most coveted demographic in television, eighteen- to thirty-four-year-olds. In years past the network's declining audience share was attributed to the siphoning off of more viewers by cable. By 2006, however, the circumstances driving the network's fall from grace were growing more ominous, due in large part to young people's migration to digital.

I mentioned to the NBC Universal researcher that I was observing some interesting behaviors in one of the most wired environments in America, college campuses. The ubiquitous presence of mobile phones, laptops, and of course the iPod in young people's lives is a constant reminder of the changes happening before our eyes. While the number of young people using the Internet and other new communication technologies is certainly notable, how the scope of their online and digital lives continues to broaden is the real story.

"Today," I said to the NBC researcher, "you have to think of media, especially digital media, as a lifestyle." So, while television still matters to young people, it is no longer the dominant media in their lives.

In March 2007, roughly six months after my initial conversation with the NBC researcher, the network's audience woes turned gloomier. According to Nielsen, Americans were abandoning television in startling numbers as more than 2.5 million fewer people were watching the "big four" compared to a year earlier. During a two-week stretch that March, NBC experienced its worst ratings in more than twenty years.

It was around this time that we began studying young people's media behaviors up close. Our immersion in the digital trenches, the spaces where young people make and remake their digital lives, was timely. According to the Pew Internet & American Life Project, 2006 was the tipping point for high-speed Internet connections, turning what Pew, two years earlier, called the "broadband elite" into the broadband masses. Furthermore, 2006 was the year that three of the most celebrated Web 2.0 brands—MySpace, Facebook, and YouTube—established a formidable presence in American popular culture. Inspired by the popular explosion of new Web brands and the commercial potential of what Net-entrepreneurs cleverly began marketing as Web 2.0, *Time* magazine named user-generated media its Person of the Year.

While much of the business news sector was celebrating the growth of the social Web, a team of researchers that I assembled began thinking about innovative ways to conduct research and fieldwork related to the rise of social and mobile media.

One of my early interviews with a high school principal established a key goal in my research. Talking about the state of schools today, she said, "When it comes to technology, we cannot be afraid to learn from our students." I never forgot her comment. It is a core principle in my own research. Throughout this project I have learned more from children, tweens, teens, and young twenty-somethings than I will ever be able to record in this book or elsewhere. In fact, just about every step of the way on my journey toward completing this book, young people have been centrally involved as either participants in the study or in the design and execution of the study. I learned from a number of young students. Two in particular: Jessica Landes and H. Erin Lee. Jessica conducted three different sets of interviews and helped my research team set up a Facebook site to interact with users of technology. Whenever Facebook made changes in the platform, she was the first on our team to know about it. Erin was an indispensable resource. The research for this book could not have been executed without her. Erin, along with a group of undergraduate students we taught, helped to build the survey instrument we continue to use. She also leads all of our statistical analysis.

Our survey was organized into four parts. Part One, "General Media," asked young people questions related to which media they own, use, and spend the most time with. Part Two of the survey addressed the "Internet" with a particular focus on how often they use the Internet, the range of activities and experiences they seek out on the Web, and degrees of happiness and satisfaction with their Web-based experiences. A subset of questions—on frequency of use, intensity of use, and attitude—focused specifically on social-network sites. "Music Media" was the central focus of Part Three. The questions in this section addressed how young people's consumption of music was evolving in the age of digital downloads, iTunes, iPods, and peer-to-peer file sharing. One of my early and enduring interests involves understanding the state and fate of television as young people migrate to digital. Part Four was designed to illuminate how often young people watch television, what platforms they prefer when watching television, as well as the attitudes they possess regarding the technology. The final section of the survey, Part Five, collected some basic personal data such as gender and race.

Erin was also the teaching assistant for three undergraduate research courses that I taught. In each of those classes we worked with students to constantly refine our questions and approach to studying digital media. The classes usually consisted of about ten to twelve students and offered a hands-on research experience that introduced students to the art and science of rigorous research. Similarly, I benefited from the experience by having a structured opportunity to engage young students about the role of technology in their lives. From these structured classes we collected more than three hundred interviews. Our young researchers were required to record and transcribe some of their interviews. They were also expected to write a summary of their interviews and a report on their findings. All of our work was situated within the context of academic research. The reports by the Pew Internet & American Life Project were quite useful too. The environment was academic but based on real-life experiences, collaborative and rigorous, and challenging but encouraging.

We did another large sample of interviews, about 105, between February and April 2008. This time our team of students focused on matters related to community online. Specifically, we wanted to learn if young people use social-network sites to bridge (defined as meeting people not like them) or to bond (defined as meeting people like them). Our survey gave us some data on this matter, but we knew the in-depth conversations would provide a different dimension of data. We sought to get an even distribution of men and women as well as a racial and ethnic cross-section. Even though these conversations were potentially sensitive, we wanted them, again, to be informal and semistructured. We began with a series of questions related to general usage of the Internet. Understanding the degree of engagement with the Web was crucial in all of our conversations. We did not plan it this way, but we ended up collecting what amounts to a life history of technology use among young people. The second set of questions focused on use of social-network sites. We wanted to learn how often and why they use social sites, which sites they prefer, and why. The final section of questions were structured around the notion of community. Specifically, we wanted to know who they connect and correspond with online and to what degree these connections are tied to their off-line affiliations. We were also interested in understanding how diverse these online networks are. Instead of imposing our own definitions of diversity, we let the students define and discuss what diverse online networks mean to them.

We distributed the survey the old-fashioned way, with pencil and paper, as well as the more modern and efficient way, online. First, we visited a wide-range of classrooms that gave us a mix of majors, years in school, and race and ethnicities. Next, we generated a random sample of students and submitted a recruitment message via e-mail, Listservs, and electronic fliers. References to the survey in this book are drawn from about 560 completed questionnaires.

From the very beginning of this project my goal was to combine a mix of methods—quantitative and qualitative. Whereas quantitative methods help researchers identify trends, frequencies, and patterns, qualitative methods provide depth, detail, and color. Quantitative data tells you, for example, that teens spend an average of three hours a day online. Qualitative data provides rich detail on the kinds of experiences generated during those three hours.

A research grant from the College of Communication Office of the Dean at the University of Texas at Austin supplied some initial support to go out into the field to conduct in-depth interviews with young technology users. In addition, Mary Celeste Kearney, Laura Stein, and Stuart Kelban—members of the Undergraduate Studies Committee—provided some funds that helped with the research. These enrichment funds allowed me to put together a small team of researchers. Alyxandra Vesey was hired as a graduate research assistant. One of our first tasks was to write a literature review of the scholarly research on social-network sites. That white paper assisted the development and design of a question guide that we used as the basis for conducting in-depth interviews with young people regarding their use of social-network sites. We also recruited two undergraduate students, Jessica Landes and Cristen Radice, to help us conduct, transcribe, and analyze the in-depth conversations. Those in-depth conversations were influenced by Herbert and Irene Rubin's book, *Qualitative Interviewing: The Art of Hearing Data*. We took their advice and made our interviews semistructured, conversational, and rigorous at the same time.

The question guide was built around three broad themes: 1) Internet use, 2) use of social-network sites, and 3) social impact of social-network sites. After conducting the interviews, each team member was then asked to write a quick summary of the key points and themes that emerged from the interviews. Then, we met to discuss the most important and common themes from the interviews. These turned out to be important learning and strategic sessions about social media, the rise and influence of social-network sites, and their everyday life implications. What I valued most about these sessions and, indeed, all of the research for this book, is that the young and the digital were integrally involved from the very beginning. Young people helped us design our survey and the questions for our in-depth conversations. Finally, each team member was asked to produce a final report that documented their interviews and analysis and also to submit the audio recordings of their interviews.

Blake Rebouche, another undergraduate student and gamer, helped to identify and interview *World of Warcraft* users.

The formation of the four-pack deserves some mention here. I met four-pack member Derrick first. During that interview we talked about growing up with computers and his use of social-network sites. As our conversation unfolded I learned a lot about Derrick, including his passion for games. I asked Derrick to identify a handful of his peers for a panel that I wanted to put together. Several young men expressed an interest before I eventually selected Brad, Trevor, and Chase, along with Derrick. Once I settled on the four-person panel, I visited with them in their residential hall. The four-pack filled out the survey and also agreed to complete media journals. Every two weeks I issued them questions via e-mail to address in their media diaries. One week, for example, the diaries may have been devoted to games, and the next week, to television. The diaries were honest, rich in detail, and provided intimate access to a group of young men who embody the rising generation of gamers. Each four-pack member submitted a total of four diaries. Each submission was followed up with a one-on-one conversation.

We also spent endless hours in the online spaces young people inhabit, looking at MySpace and Facebook profiles, YouTube videos and comments, and blogs. Our first goal was to always keep the identities of the young people we observed and the schools and residences we visited anonymous. All of the names of students, principals, and teachers, for example, are pseudonyms. Also, we knew that pictures would add another dimension of detail and depth to our work. So we recruited students to collect pictures of their peers' apartments and dorm rooms in order to give us a fuller sense of the role, placement, and impact of technology on their environment.

The Young and the Digital was greatly enhanced by the generosity of a number of schools, teachers, and principals who welcomed me into their complex world. In order to understand young people's engagement with technology, I knew I had to get into schools, places that are like no other when it comes to understanding young people. Observing second and third graders in class or talking with a group of seventh and eighth graders about MySpace, Facebook, or what it means to be a citizen in the digital world is always illuminating and rejuvenating.

My participation in a technology task force at one Austin-based private school has been a constant source of knowledge sharing, idea creation, and learning. Our meetings dealt primarily with what technology means for the learning environment that this particular school establishes for its students. Our conversations maintained an unswerving focus on two questions: First, how is technology going to enhance the student's learning experience? Second, how does technology impact the community aspects of the school? The latter question relates to cyberbullying, digital citizenship, and the social and communal aspects of technology. It is one thing to read about these matters and another one to work with a school to learn and set relevant policies and boundaries that empower students' engagement with technology.

In 2006 I was selected to join the fabulous team that the MacArthur Foundation put together to push its young people and digital initiative forward. Specifically, I want to thank Anna Everett and Katie Salen. My primary involvement was collaborating with a group of scholars and game designers charged by MacArthur to map out why and how games matter in the lives of young people. The chapter Everett and I coauthored appears in the book *The Ecology of Games: Connecting Youth, Games, and Learning.*

Some of the ideas for this book have been cultivated in various talks and presentations made at universities, health groups, and foundations. I want to thank Henry Jenkins for an invitation to speak and help kick off MIT's Media in Transition conference in 2007. A couple of visits to Stanford to talk about race and new media helped me to refine many of my ideas. Howard Winant and Melvin Oliver's invitation to speak at the University of California at Santa Barbara was a nice opportunity to meet graduate students interested in media and sociology. Likewise, meeting students and faculty in the Media, Society, and Technology Program at Northwestern University sharpened some of my ideas about race, social-network sites, and digital gates. Visits to the University of Michigan and Harvard also gave me an opportunity to talk with colleagues about the shifting landscape of new media and its relationship to young people. I am forever grateful for the invitations to speak with the Lance Armstrong Foundation Young Adult Alliance and Cleveland Clinic Children's Hospital about youth and new media. Sharon Strover's invitation to lecture and teach workshops on youth and digital media culture at the New University of Lisbon and Porto University has connected me to an interesting community of scholars and professionals from Portugal who are beginning to grapple with the global migration to digital.

This book has taken me on an amazing ride. I built community and shared conversations with so many people along the way that I cannot even come close to naming them all. Also, we collected so much data from interviews and the surveys that there is no way all of it could be included in the book. Therefore we have decided to start a blog as a way to keep our work alive, public, and involved in the conversations of which you are undoubtedly a part.

As always there are so many people to thank. Many of my colleagues at the University of Texas at Austin offered various forms of support, including Jacqueline Vickers, Rod Hart, Joe Straubhaar, Susan Dirks, Christine Williams, Gloria Holder, Bert Hergistad, Tom Schatz, Kathleen Tyner, and Karin Wilkins.

Many other scholars and researchers also provide intellectual community: Herman Gray, Susan Douglas, Robin Means-Coleman, Amanda Lotz, Scott Campbell, Jeff Chang, Mark Anthony Neal, E. Patrick Johnson, Barbara and Daniel O'Keefe, Noshir Contractor, Jennifer Light, Sean Zehnder, Ellen Wartella, Dawn Elissa-Fischer, Marceleyna Morgan, Harry Elam, Michele Elam, and Orlando Patterson.

My conversations with friends and teachers at Trinity Episcopal School have been a constant source of nourishment. Thank you all. Lisa Zapalac, Brenda Leaks, Karen Calvert, Donna Hallet, Rebecca McClure, Jeff Pzynski, Ron Olfers, Ben Davis, Sandy Cangelosi, Terri and Chris Von Dohlen, Jeff McMahon, Bruce Ezell, Claire Saunders, Juliet Tapia, and Liza Lee shared their thoughts, helpful books, or interests throughout my research.

Great educators and community leaders like Mark Cunningham, Sherry Watkins, Rana Emerson, Holly Custard, Malinda McCormick, and Dale Thompson were incredible resources.

As usual, my friends at Beacon Press have been outstanding, encouraging, and enthusiastic champions of this book. From the moment we talked about my ideas for this project one evening in a Boston bar, my editor Gayatri Patnaik has offered steady guidance and unswerving confidence. Despite her busy schedule, she read a very rough draft and saw promise. Likewise Joanna Green has been a great resource. Also, thanks to Tom Hallock, Sarah Laxton, Susan Lumenello, Pamela MacColl, and Caitlin Meyer for their support.

Now that I am finished with this book I may actually find the time to use Facebook and "friend" my friends.

Finally, my family has been central in writing this book. Though she is not a digital native by any stretch of the imagination, my mother, Jeweline Watkins, has forced me to think about the wider public while writing this book. The memory of my father is never far away. My wife deserves special thanks for keeping our family ready for the world. Her love, energy, and embrace is a constant source of life. My daughter, Cameron Grace, is the light of my world and her third-grade basketball and volleyball teams rock too!

INTRODUCTION: **The Young and the Digital**

ix "'I didn't do as much'": Rupert Murdoch, "Speech by Rupert Murdoch to the American Society of Newspaper Editors" (lecture, Marriott Hotel, Washington, DC, April 13, 2005), www.newscorp.com/news/news_247.html.

ix "'We may never become'": Ibid.

x "'It is a monumental'": Ibid.

x *Abandoning the News*": Merrill Brown, "What's the Future of the News Business? This Report to Carnegie Corporation of New York Offers Some Provocative Ideas," *Carnegie Reporter* 3, no. 2 (Spring 2005).

xi "In 2005 the fastest growing Web sites": Nielsen//NetRatings, "Nielsen//NetRatings Reports the Fastest Growing Web Sites Year-over-Year Among Top Internet Properties: Apple, Google and Amazon Take the Lead," news release, December 20, 2005.

xi "Between November 2004": Ibid.

xi "During the same period": Ibid.

xi "For the period of": Andrew Lipsman, *More than Half of MySpace Visitors Are Now Age 35 or Older, as the Site's Demographic Composition Continues to Shift* (Reston, VA: comScore Media Metrix, October 5, 2006), www.comscore.com/press/release.asp?press=1019.

xii "'Since we were telling'": Alex Williams, "Do You MySpace?" *New York Times*, August 28, 2005, ST1.

xii " 'People are starting to understand' ": Antony Bruno, "MySpace Is the (Online) Place," *Billboard*, July 2, 2005.

xii " 'The Internet is exciting again' ": John Battelle, "Building a Better Boom," *New York Times*, November 18, 2005.

xviii " 'TV watching comes at the expense' ": Robert Putnam, *Bowling Alone: The Collapse and Revival of American Community* (New York: Simon & Schuster, 2000), 237.

xix " 'Increasingly, her citizens are encouraged' ": Ray Oldenburg, *The Great Good Place: Cafes, Coffee Shops, Bookstores, Bars, Hair Salons, and Other Hangouts at the Heart of a Community* (New York: Paragon House, 1989), xvi.

xx "Just as separation in": Marshall Van Alstyne and Erik Brynjolfsson, "Electronic Communities: Global Village or Cyberbalkans?" (working paper, Sloan School, Massachusetts Institute of Technology, March 1997), 3, http://web.mit.edu/marshall/www/papers/CyberBalkans.pdf.

xx "In 2000 the number of households": U.S. Bureau of the Census, *Home Computers and Internet Use in the United States: August 2000*, prepared by the Department of Commerce in cooperation with the Economics and Statistics Administration, Bureau of the Census, Washington, DC, 2001.

ONE: Digital Migration

1 " 'The enduring American love affairs' ": Otto Friedrich, "The Computer," *Time*, January 4, 1983.

1 "Between 1980 and 1990 multi-television set": Nielsen, *Television Audience: 1993*, Nielsen Media Research, 2.

2 "During that same period": Ibid., 2, 9.

2 "In 1980 few Americans": Ibid., 2.

2 "For the first time in": Donald F. Roberts and Ulla G. Foehr, *Kids and Media in America* (New York: Cambridge University Press, 2003).

2 "Since the 1990s American homes": See Jorge Reina Schement, "Wiring the Castle: Demography, Technology, and the Transformation of the American Home" (lecture, MIT Communications Forum: Media in Transition, Cambridge, MA, March 2006).

2 "In 1985, when the": U.S. Bureau of the Census, *Home Computers and Internet Use in the United States*.

2 "By comparison, 98 percent": Nielsen, *Television Audience: 1993*, 2.

3 "In 2003, according to the U.S. Census Bureau": U.S. Bureau of the Census, *Computer and Internet Use in the United States: 2003*, prepared by the Department of Commerce in cooperation with the Economics and Statistics Administration, Bureau of the Census, Washington, DC, 2005.

3 "The diffusion of the Internet": Ibid., 3.

3 "In 2003, 76 percent of homes": Ibid., 2.

3 "Also, homes with school-age children": Ibid., 3.

5 "'Starting junior high seems to be'": Amanda Lenhart, Mary Madden, and Paul Hitlin, *Teens and Technology* (Washington, DC: Pew Internet & American Life Project, July 27, 2005): 2.

5 "In 2005 about 60 percent": Ibid.

6 "'IM satisfies two major'": Bonka Boneva et al., "Teenage Communication in the Instant Messaging Era," in *Computers, Phones, and the Internet: Domesticating Information Technology*, eds. Robert E. Kraut, Malcolm Brynin, and Sara Kiesler (Cambridge, UK: Oxford University Press, 2006), 241.

6 Discussion of the initial reporting from the Carnegie Mellon HomeNet field trial is reported by Robert Kraut et al., "The HomeNet Field Trial of Residential Internet Services," *Communications of the ACM* 39, no. 12 (December 1996): 55–63.

7 "'Teenagers' enthusiasm motivated other family'": Ibid., 57.

7 "'became the technical support gurus'": David Frohlich and Robert Kraut, "The Social Context of Home Computing," in *Inside the Smart Home*, ed. Richard Harper (New York: Springer, 2003), 155.

8 "'The availability of a broadband connection'": John B. Horrigan and Lee Rainie, *The Broadband Difference* (Washington, DC: Pew Internet & American Life Project, June 23, 2002): 14.

8 "Between March 2005 and March 2006": John B. Horrigan, *Home Broadband Adoption 2006* (Washington, DC: Pew Internet & American Life Project, May 28, 2006): i.

8 "During that same period": Ibid., 3.

8 "'Nothing else in the twentieth century'": Putnam, *Bowling Alone*, 221.

8 "'we spent almost all six'": Ibid., 222–23.

9 "Many of the earliest studies": See, for example, Robert Kraut et al., "Examining the Effects of Internet Use on Television Viewing: Details Make a Difference," in *Computers, Phones, and the Internet*; Nielsen//NetRatings, "TV Viewing in Internet Households," A Report by Nielsen Media Research, May 1999.

9 "'Internet homes are lighter TV viewers'": Nielsen//NetRatings, "TV Viewing in Internet Households." A Report by Nielsen Media Research (New York: Nielsen Media Research, May 1999).

10 "That year Pew reported that": Horrigan and Rainie, *Broadband Difference*, 24.

10 "Three things distinguished broadband users": Ibid., 10–14.

10 " 'The increased daily usage translates' ": Ibid., 13.

10 "That same 2002 study": Ibid., 12.

10 "The difference in media streaming": Ibid.

10 " 'decrease in television viewing' " Ibid., 24.

10 "By 2004 more than half": Amanda Lenhart and Mary Madden, *Teen Content Creators and Consumers* (Washington, DC: Pew Internet & American Life Project, November 2, 2005): 1.

11 "Pew has a name": Amanda Lenhart, John Horrigan, and Deborah Fallows, *Content Creators Online* (Washington, DC: Pew Internet & American Life Project, February 29, 2004): 8.

13 "TV is controlled by adults": Don Tapscott, *Growing Up Digital: The Rise of the Net Generation* (New York: McGraw-Hill, 1997): 25.

13 "In 1998, just as the network computer": Ibid, 25–31.

13 "So much has changed": Ben Grossman, "Jeff Zucker: After Landing Leno, Now What?" *Broadcasting & Cable*, December 14, 2008, http://www .broadcastingcable.com/article/160735-Jeff_Zucker_After_Landing_Leno_ Now_What_.php?rssid=20100&q=jeff+zucker.

13 "We are at a critical point": Ibid.

13 All of the data on virtual worlds and kids discussed here is reported by LeeAnn Prescott, "Virtual Worlds Ranking—Runescape #1," April 30, 2007, http://weblogs.hitwise.com/leeann-prescott/2007/04/virtual_worlds_ranking_ runesca.html.

15 "First of all, they report": All of the findings discussed here are reported by Victoria Rideout, Elizabeth Vandewater, and Ellen Wartella, *Zero to Six: Electronic Media in the Lives of Infants, Toddlers, and Preschoolers* (Menlo Park, CA: The Kaiser Family Foundation, fall 2003).

15 " 'Those who use a computer' ": Ibid., 5.

15 " 'a series of studies to learn' ": The Roper Organization, *America's Watching: Public Attitudes toward Television* (New York: Television Information Office, 1995): i.

15 " 'Positive feelings about television' ": Ibid., i.

16 " 'The three words,' Roper declared": Ibid., 1.

16 "Significantly, 65 percent": Paul Taylor, Cary Funk, and April Clark, *Luxury or Necessity? Things We Can't Live Without: The List Has Grown in the Past Decade* (Washington, DC: Pew Research Center, December 14, 2006): 4.

TWO: Social Media 101

20 "is offered by a commercial entity": *Deleting Online Predators Act of 2006*, HR 5319, 109th Cong., 2nd sess. (July 26, 2006): H 5883–5889, www.ala.org/ ala/aboutala/offices/wo/woissues/techinttele/dopa/HouseDOPADebate.pdf.

20 "'Social networking sites, best known'": Declan McCullagh, "Chat Rooms Could Face Expulsion," July 27, 2006, http://news.cnet.com/2100-1028_3-6099414.html.

20 "'have become a haven'": Ibid.

20 "'When children leave the home'": Declan McCullagh, "Lawmakers Take Aim at Social-Networking Sites," May 11, 2006, http://news.cnet.com/ Congress-targets-social-network-sites/2100–1028_3–6071040.html.

20 "By 2004 a decisive majority": Lenhart, Madden, and Hitlin, *Teens and Technology*, 1.

20 "According to the Pew Internet & American Life Project": Amanda Lenhart and Mary Madden, *Social Networking Websites and Teens: An Overview* (Washington, DC: Pew Internet & American Life Project, 2007): 1.

21 "affirms the importance of": www.ala.org/ala/aboutala/offices/oif/if issues/onlinesocialnetworks.pdf.

22 "MySpace may have been the brand": Lipsman, *More than Half of MySpace Visitors Are Now Age 35 or Older*.

22 "One in seven youth are exposed": Janis Wolak, Kimberly Mitchell, and David Finkelhor, *Online Victimization of Youth: Five Years Later* (Alexandria, VA: National Center for Missing & Exploited Children, 2006).

23 "'Even more startling'": *Deleting Online Predators Act of 2006*.

23 "'hunting predators is both the coolest'": Vanessa Grigoriadis, "'To Catch a Predator': The New American Witch Hunt for Dangerous Pedophiles," *Rolling Stone*, July 30, 2007.

23 "A 2007 report by the Pew Internet & American Life Project": Lenhart and Madden, *Social Networking Websites and Teens*.

24 "'new media forms have altered'": The John D. and Catherine T. Mac-Arthur Foundation, *Living and Learning with New Media: Summary of Findings from the Digital Youth Project*, Report on Digital Media and Learning, November 2008, 2.

24 "'Given the popularity and reach'": Beth Evans, "Your Space or My-Space?" *netConnect*, October 15, 2006, 8–12.

25 "A 2005 study of teens": Lenhart, Madden, and Hitlin, *Teens and Technology*, 14.

30 "'In our schools, every classroom'": President William Jefferson Clin-

ton, "State of the Union Address" (U.S. Capitol, Washington, DC, January 23, 1996), http://clinton2.nara.gov/WH/New/other/sotu.html.

30   " 'every classroom and every library' ": Ibid.

30   " 'in essence, information 'have nots' ' ": U.S. Department of Commerce, *Falling through the Net: A Survey of the 'Have Nots' in Rural and Urban America* (Washington, DC: National Telecommunications and Information Administration, July 1995), www.ntia.doc.gov/ntiahome/fallingthru.html.

31   " 'the romance with the machine' ": Amy Harmon, "Internet's Value in U.S. Schools Still in Question," *New York Times*, October 25, 1997.

31   "The U.S. Department of Education": Anne Kleiner and Laurie Lewis, *Internet Access in U.S. Public Schools and Classrooms: 1994–2002* (Washington, DC: U.S. Department of Education, October 2003): 4.

31   "whereas 89 percent of college-educated Latinos": The figures regarding Latino online behaviors can be found in Susannah Fox and Gretchen Livingstone, *Latinos Online* (Washington, DC: Pew Internet & American Life Project, March 14, 2007).

32   "For instance, in a 2005 survey": Lenhart, Madden, and Hitlin, *Teens and Technology*, 1.

32   "Statistical analysis of computer": Matthew DeBell and Chris Chapman, *Computer and Internet Use by Students in 2003: Statistical Analysis Report* (Washington, DC: U.S. Department of Education, Institute of Education Sciences, September 2006).

33   "When Pew conducted its": Lenhart and Madden, *Social Networking Websites and Teens*, 3.

34   " 'Older girls [15–17] in particular' ": Ibid., 2.

36   "A 2007 report by": Amanda Lenhart, *Cyberbullying and Online Teens* (Washington, DC: Pew Internet & American Life Project, June 27, 2007): 1.

36   " '41% of today's teens' ": Amanda Lenhart, *Teens, Privacy & Online Social Networks* (Washington, DC: Pew Internet & American Life Project, April 18, 2007): 7.

38   " 'there is a misconception' ": Andrew Lipsman, *More than Half of My-Space Visitors Are Now Age 35 or Older, as the Site's Demographic Composition Continues to Shift.*

38   "The percentage of people using Facebook": Ibid.

39   " 'Why are you holding out on' ": Association of Texas Professional Educators, "Use Caution in a Place Like MySpace," www.atpe.org/protection/employmentrights/mySpaceDangers.asp.

40   " 'avoid adding students to your 'friend' ' ": Ibid.

40 "'could prompt allegations of misconduct'": Ibid.

43 "A study published in": Elisheva F. Gross, "Adolescent Internet Use: What We Expect, What Teens Report," *Journal of Applied Developmental Psychology* 25 (2004): 633–49.

44 "And yet an image sent": Chana Joffe-Walt, "'Sexting': A Disturbing New Teen Trend," March 11, 2009, www.npr.org/templates/story/story.php?storyId=101735230.

44 "A sexting incident in a Pennsylvania high school": Mike Brunker, "'Sexting' Surprise: Teens Face Child Porn Charges," January 15, 2009, www.msnbc.msn.com/id/28679588/.

44 "A 2005 study found": Dale Kunkel et al., *Sex On TV 4: A Kaiser Family Foundation Report* (Menlo Park, CA: The Henry J. Kaiser Family Foundation, November 2005).

45 "'in order to cooperate with more people'": Howard Rheingold, *Smart Mobs: The Next Social Revolution* (New York: Basic Books, 2002), xxi.

THREE: The Very Well Connected

48 "Some researchers refer to it as": Mizuko Ito and Daisuke Okabe, "Technosocial Situations: Emergent Structuring of Mobile E-mail Use," in *Personal, Portable, Pedestrian: Mobile Phones in Japanese Life*, eds. Ito, Okabe, and Misa Matsuda (Cambridge, MA: MIT Press, 2005), 257–73. The authors explain that mobile phones create "new kinds of bounded places that merge the infrastructures of geography and technology, as well as technosocial practices that merge technical standards and social norms."

48 "This is called 'presence-in-absence'": M. Lombard and T. Ditton, "At the Heart of It All: The Concept of Presence," *Journal of Computer-Mediated Communication* 3, no. 2 (September 1997). The authors write that presence-in-absence or presence "is the extent to which a medium is perceived as sociable, warm, sensitive, personal or intimate when it is used to interact with other people" but also means the sense of "being there" or being aware of another's presence, or when mutual, "co-presence."

49 "'conversation is an important social process'": Claude Fischer, *America Calling: A Social History of the Telephone to 1940* (Berkeley: University of California Press, 1994), 231.

49 "'elevators and rubber tires'": Howard Rheingold, *The Virtual Community: Homesteading on the Electronic Frontier* (Cambridge, MA: MIT Press, 2000), 330.

50 "According to a 2006 study": Ulla G. Foehr, *Media Multitasking Among Youth: Prevalence, Predictors, and Pairings* (Menlo Park, CA: The Henry J. Kaiser Family Foundation, 2006).

50 "As they observe their parents'": Matt Richtel and Brad Stone, "For Toddlers, Toy of Choice is Tech Device," *New York Times*, November 29, 2007.

50 "A growing number of health professionals": Centers for Disease Control and Prevention, "Overweight Among U.S. Children and Adolescents," National Health and Nutrition Examination Survey, 2002, www.cdc.gov/nchs/data/nhanes/databriefs/overwght.pdf; S. Gortmaker et al., "Television Viewing as a Cause of Increasing Obesity among Children in the United States, 1986–1990," *Archives of Pediatrics & Adolescent Medicine* 150 (April 1996): 356–62; Elizabeth A. Vandewater, Mi-suk Shim, and Allison G. Caplovitz, "Linking Obesity and Activity Level with Children's Television and Video Game Use," *Journal of Adolescence* 27, no. 1 (February 2004): 71–85.

52 "In 1933 researchers Malcolm Wiley": Malcolm M. Wiley and Stuart A. Rice, *Communication Agencies and Social Life* (New York: McGraw-Hill, 1933).

52 "'of those values that inhere'": Ibid., 240.

52 The "far-flung relationships" reference comes from Ron Westrum, *Technologies and Society: The Shaping of People and Things* (Belmont, CA: Wadsworth, 1991). The reference to "inauthentic intimacy" comes from Fischer, *America Calling*.

52 "New communication technologies also": Joshua Meyrowitz, *No Sense of Place: The Impact of Electronic Media on Social Behavior* (New York: Oxford University Press, 1985).

53 "'The idea of a community'": Rheingold, *Virtual Community*, xv.

53 "'In traditional kinds of communities'": Ibid., 11.

54 "The telephone began as a novelty": Fischer, *America Calling*, 23.

55 "'The number of total contacts'": Ibid., 258.

56 "'seemingly obsessive need to connect'": Betsy Israel, "The Overconnecteds," *New York Times*, November 5, 2006.

56 "Seventy-five percent of American teens": Lenhart, Madden, and Hitlin, *Teens and Technology*, 15.

56 "That same report found": Ibid., 16.

57 "Among other things, teens use IM": Bonka Boneva et al., "Teenage Communication in the Instant Messaging Era."

57 "Teens and young twenty-somethings": *Consumers in the 18-to-24 Age Segment View Cell Phones as Multi-Functional Accessories; Crave Advanced Features and Personalization Options* (Reston, VA: comScore Media Metrix, January 22, 2007), www.comscore.com/press/release.asp?press=1184.

57 "A 2005 report by NOP World Technology mKids": Brand Noise, "NOP Study: Nearly Half of US Teens and Tweens Have Cell Phones," March 10, 2005, http://brandnoise.typepad.com/brand_noise/2005/03/nop_study_nearl.html.

57 "What's more, a 2005 report": Lee Rainie and Scott Keeter, *How Americans Use Their Cell Phones* (Washington, DC: Pew Internet & American Life Project, April 3, 2006): 6.

57 "'for a teenager to send'": Margaret Web Pressler, "For Texting Teens, an OMG Moment When the Phone Bill Arrives," *Washington Post*, May 20, 2007.

57 "The 2008 CTIA Semi-Annual Wireless Industry Survey": "CTIA— The Wireless Association® Releases Latest Wireless Industry Survey Results," September 10, 2008, www.ctia.org/media/press/body.cfm/prid/1772.

58 "[It] has the personality of": Williams, "Do You MySpace?"

58 "Forget the mall": Janet Kornblum, "Teens Hang Out at MySpace," *USA Today*, January 8, 2006.

58 "the biggest mall-cum-nightclub-cum-7-Eleven": Spencer Reiss, "His Space," *Wired*, July 2006.

58 "'a great variety of public places'": Ray Oldenburg, *The Great Good Place*, 16.

59 "'The benefits of participation'": Ibid., 43.

59 "In his classic study": Dick Hebdige, *Subculture: The Meaning of Style* (London: Methuen, 1979).

63 "Many of the first studies": Robert Kraut et al., "Internet Paradox: A Social Technology That Reduced Social Involvement and Psychological Well-Being?" *American Psychologist* 53, no. 9 (1998): 1017–31.

63 "Some researchers found evidence": I. Shklovski, R. E. Kraut, and L. Rainie, "The Internet and Social Participation: Contrasting Cross-Sectional and Longitudinal Analyses," *Journal of Computer-Mediated Communication* 10, no. 1 (November 2004): http://jcmc.indiana.edu/vol10/issue1/shklovski_kraut.html.

63 "'one simply can not be engaged'": Norman H. Nie, D. Sunshine Hillygus, and Lutz Erbing, "Internet Use, Interpersonal Relations and Sociability: Findings from a Detailed Time Diary Study," in *The Internet in Everyday Life*, eds. Barry Wellman and Caroline Haythornthwaite (Oxford, UK: Blackwell Publishers, 2003).

64 "'brought neighbors together to socialize'": Keith N. Hampton and Barry Wellman, "Netville On-line and Off-line: Observing and Surveying a Wired Suburb," *American Behavioral Scientist* 43, no. 3 (November 1999): 492.

64 "In fact, young, college-educated": See Shaila Dewan, "Cities Compete in Hipness Battle to Attract Young," *New York Times*, November 25, 2006.

65 "A 2007 report from": Lenhart and Madden, *Social Networking Websites and Teens*, 2.

65 "Pew also found that": Ibid., 2.

68 "Social-network sites, Ellison, Steinfield, and Lampe": Nicole B. Ellison, Charles Steinfield, and Cliff Lampe, "The Benefits of Facebook 'Friends': Social Capital and College Students' Use of Online Social Network Sites," *Journal of Computer-Mediated Communication* 12 (2007): 1143–68.

68 "Highly engaged users are": Ibid.

68 "Additionally, the ability to stay": Ibid.

68 "Sociologist Mark Granovetter's work shows": Mark S. Granovetter, "The Strength of Weak Ties," *American Journal of Sociology* 78, no. 6 (1973): 1360–80. Also, see Granovetter, "The Strength of Weak Ties: A Network Theory Revisited," in *Social Structure and Network Analysis*, eds. P. V. Mardsen and N. Lin (Thousand Oaks, CA: Sage Publications, 1982), 105–30.

70 "more time at each site": Nielsen//NetRatings, "Teens Who Visit Both MySpace and Facebook Drive Time Spent at the Social Networking Sites, According to Nielsen//NetRatings," September 20, 2007, www.nielsen-online .com/pr/pr_070920.pdf.

71 "What Chase described as": Frank Vetere, Steve Howard, and Martin R. Gibbs, "Phatic Technologies: Sustaining Sociability through Ubiquitous Computing," in *Proceedings of ACM CHI 2005 Conference on Human Factors in Computing Systems 2005* (Workshop) (Portland, Oregon: April 2–7, 2005).

71 "'to keep channels of communication open'": Ibid.

72 "'the facility to idly chat'": Ibid.

72 "'It was the reassurance'": Ibid.

72 "'laden with emotional significance'": Ibid.

72 "'maintain and strengthen existing relationships'": Ibid.

72 "It is widely recognized": James E. Katz and Mark Aakhus, eds., *Perpetual Contact: Mobile Communication, Private Talk, Public Performance* (Cambridge, UK: Cambridge University Press, 2002).

FOUR: Digital Gates

75 "'Americans aren't so good at talking'": dana boyd, "Viewing American Class Divisions through Facebook and MySpace," June 24, 2007, www.zephoria .org/thoughts/archives/2007/06/24/viewing_america.html.

76 "'Facebook kids,' the blogger writes": Ibid.

76 "kids whose parents didn't go to college": Ibid.

78 "'Hispanic students,' she writes": Eszter Hargittai, "Whose Space? Differences among Users and Non-Users of Social Network Sites," *Journal of Computer-Mediated Communication* 13, no. 1 (2007), http://jcmc.indiana.edu/vol13/issue1/hargittai.html.

78 "Children growing up in low": Donald F. Roberts et al., *Kids & Media & the New Millennium* (Menlo Park, CA: The Henry J. Kaiser Family Foundation, 1999).

78 "'The most pronounced finding'": Hargittai, "Whose Space?"

82 "Bourdieu made a career": Pierre Bourdieu, *Distinction: A Social Critique of the Judgment of Taste* (Cambridge, MA: Harvard University Press, 1984).

86 "For eight years, Setha Low": Setha Low, *Behind the Gates: Life, Security, and the Pursuit of Happiness in Fortress America* (New York: Routledge, 2003).

86 "Analysis of a 1997 national survey": Edward J. Blakely and Mary Gail Snyder, *Fortress America: Gated Communities in the United States* (Washington, DC: Brooking Institute, 1997).

86 "'primarily about stability and a need'": Low, *Behind the Gates*, 23.

86 "desire for safety, security, community": Ibid., 9.

87 "'the more purified the environment'": Ibid., 143.

90 "In the span of two months": Yuki Noguchi, "In Teens' Web World, MySpace Is So Last Year," *Washington Post*, October 29, 2006.

91 "Ten months after opening": Nielsen//NetRatings, "Teens Who Visit Both MySpace and Facebook."

91 "In December 2006 Facebook": These statistics are reported by Facebook, www.facebook.com/home.php?#/press/info.php?timeline.

93 "In a 2007 report titled *Latinos Online*": Susannah Fox and Gretchen Livingstone, *Latinos Online* (Washington, DC: Pew Internet & American Life Project, March 14, 2007).

95 "'some forms of capital'": Robert Putnam, *Bowling Alone*, 22.

95 "'outward looking and encompasses'": Ibid.

98 "'We have come to expect'": Bill Bishop, *The Big Sort: Why the Clustering of Like-Minded America is Tearing Us Apart* (Boston: Houghton Mifflin, 2008), 213.

98 "'Kids have grown up in'": Ibid.

98 "'the most outward-looking . . . generation'": John Zogby, *The Way We'll Be: The Zogby Report on the Transformation of the American Dream* (New York: Random House, 2008), 115.

100 "In one of the first 'virtual-field studies'": Paul W. Eastwick and Wendi

L. Gardner, "Is It a Game? Evidence for Social Influence in the Virtual World," *Social Influence* 4, no. 1 (January 2009): 18–32.

101 "the virtual world may not prove": Ibid., 29.

FIVE: **We Play**

103 "Compared to the year before": Bill Carter, "The Absence of Television Viewers Has Network Executives Scratching Their Heads," *New York Times*, October 22, 2003, c8.

103 "'a nuclear strike, a smallpox outbreak'": John Schwartz, "Leisure Pursuits of Today's Young Man," *New York Times*, March 29, 2004.

103 "'Frankly what we're seeing strains credulity'": Carter, "The Absence of Television Viewers."

104 "'should have been seen'": Schwartz, "Leisure Pursuits."

104 "'You never see these kind'": Cynthia Littleton and Andrew Wallenstein, "Nielsen Study Helps Explain Demo Declines," *The Hollywood Reporter*, November 26, 2003.

104 "Young people still watch television": Trip Gabriel, "Decoding What 'Screen-Agers' Think About TV," *New York Times*, November 25, 1996.

104 "'The fact that more than'": *Desperately Seeking Men Aged 18–34? Find Them Online, Says comScore Media Metrix* (Reston, VA: comScore Media Metrix, November 4, 2003), http://www.comscore.com/press/release.asp?press=361.

105 "In a 2007 report titled": Mary Madden, *Online Video* (Washington, DC: Pew Internet & American Life Project, July 25, 2007): 9.

105 "Between 2004 and 2006 there was": Nielsen Wireless and Interactive Services, "The State of the Console: Video Game Console Usage Fourth Quarter 2006," 2007.

105 "'U.S. computer and video game software'": Entertainment Software Association, "Industry Facts," www.theesa.com/facts/index.asp.

105 "A study by Nielsen Entertainment": Nielsen Entertainment, "Nielsen Entertainment Study Shows Video Gaming is Increasingly a Social Experience," news release, October 5, 2006, //www.prnewswire.com/cgi-bin/stories .pl?ACCT=104&STORY=/www/story/10-05-2006/0004446115&EDATE.

105 "'as games continue to increase'": Nielsen Entertainment, "Nielsen Entertainment Reports on Mobile and Video Game Entertainment in Today's Evolving Consumption Landscape in Two Separate Studies to Publish Tomorrow," news release, November 21, 2005.

107 "Ultimately, no matter if it is the promise": Katie Salen and Eric Zimmerman, *Rules of Play: Game Design Fundamentals* (Cambridge, MA: MIT Press, 2003).

107 "In its third annual": Nielsen Entertainment, "Nielsen Entertainment Study," //www.prnewswire.com/cgi-bin/stories.pl?ACCT=104&STORY=/www/story/10-05-2006/0004446115&EDATE.

110 "Nine of the ten top-selling games": Also, see The NPD Group, "Playing Video Games Viewed As Family/Group Activity and Stress Reducer," December 12, 2007, www.npd.com/press/releases/press_071212.html.

110 "'It's a very interesting and frustrating thing'": Seth Schiesel, "As Gaming Turns Social, Industry Shifts Strategies," *New York Times*, February 28, 2008.

110 "'And our Japanese colleagues'": Ibid.

110 "'feels like a brawny but'": Seth Schiesel, "A Weekend Full of Quality Time with PlayStation 3," *New York Times*, November 20, 2006.

110 "'for many consumers, it's still'": Martin Fackler, "Hobbled by Disappointing Sales and a Loss at the Game Unit, Sony's Profit Drops 5%," *New York Times*, January 31, 2007.

111 "It is easy to forget": Dmitri Williams, "A Brief Social History of Game Play," in *Playing Video Games: Motives, Responses, and Consequences*, eds. Peter Vorderer and Jennings Bryant (New York: Routledge, 2006).

111 "'the Wii is meant to'": Seth Schiesel, "Getting Everybody Back in the Game," *New York Times*, November 24, 2006.

111 "'In an entirely counterintuitive'": Schiesel, "As Gaming Turns Social."

111 "'realized that emphasizing the communal'": Ibid.

113 "'crafted places,' he writes, 'inside computers'": Edward Castronova, *Synthetic Worlds: The Business and Culture of Online Games* (Chicago: University of Chicago Press, 2005), 4.

113 "In a 2005 large-scale survey": Nicholas Yee, "The Psychology of Massively Multi-User Online Role-Playing Games: Motivations, Emotional Investment, Relationships and Problematic Usage," in *Avatars at Work and Play: Collaboration and Interaction in Shared Virtual Environments*, eds. Ralph Schroder and Ann-Sofie Axelsson (London: Springer, 2006), 193.

113 "respondents had spent at least": Ibid.

113 "Eight percent of users": Ibid.

114 "'two games in one'": For a good discussion of this phenomenon, see, for example, Nicolas Ducheneaut et al., "'Alone Together?' Exploring the Social Dynamics of Massively Multiplayer Online Games," in *Proceedings of ACM CHI 2006 Conference on Human Factors in Computing Systems 2006* (New York: ACM Press, 2006): 407–16. Also, see Ducheneaut et al., "Building an MMO with Mass Appeal: A Look at Gameplay in World of Warcraft," *Games and Culture* 1, no. 4 (October 2006): 281–317.

114 "Kurt Squire, a games scholar": Kurt Squire, "Open-Ended Video Games: A Model for Developing Learning for the Interactive Age," in *The Ecology of Games: Connecting Youth, Games, and Learning*, ed. Katie Salen (Cambridge, MA: MIT Press, 2008).

115 "'Story,' the authors write": Edward F. Schneider et al., "Death with a Story: How Story Impacts Emotional, Motivational, and Physiological Responses to First-Person Shooter Video Games," *Human Communication Research* 30, no. 3 (July 2004): 361–75.

115 "Despite the 'cult of the amateur'": For a critique of user-generated media, see Andrew Keen, *The Cult of the Amateur: How Today's Internet Is Killing Our Culture* (New York: Doubleday Business, 2007).

116 "'I like that I can be somebody else'": Yee, "The Psychology of Massively Multi-User Online Role-Playing Games," 194.

116 "Researchers call this gender swapping": Sherry Turkle, *Life on the Screen: Identity in the Age of the Internet* (New York: Simon & Schuster, 1997), 212–26.

116 "'I became the biggest black guy'": Ketzel Levine, "Alter Egos in a Virtual World," *Morning Edition*, National Public Radio, July 31, 2007, www.npr.org/templates/story/story.php?storyId=12263532.

117 "'I wanted to know more about'": Turkle, *Life on the Screen*, 216.

117 "'As a man I was'": Ibid.

117 "In most cases, critics point out": For a discussion on the complexities of gender identity in the online world, see Jodi O'Brien, "Writing in the Body: Gender (Re) Production in Online Interaction," in *Communities in Cyberspace*, eds. Peter Kollock and Marc A. Smith (New York: Routledge, 1999). Also, see *Fair Play? Violence, Gender and Race in Video Games* (Oakland, CA: Children Now, 2001) for one of the first content analyses of gender representation in video games.

119 "Yee and Bailenson conducted two experiments": The discussion of Nicholas Yee and Jeremy Bailenson's test of the "Proteus Effect" is drawn from Yee and Bailenson, "The Proteus Effect: The Effect of Transformed Self-Representation on Behavior," *Human Communication Research* 33 (2007): 271–90.

120 "'These two studies,' Yee and Bailenson write": Ibid.

121 "'10% of [MMORPG] users felt'": Yee, "The Psychology of Massively Multi-User Online Role-Playing Games," 201.

121 "'the process of becoming'": John Seely Brown and Douglas Thomas, "You Play World of Warcraft? You're Hired!" *Wired*, April 2006.

124 "In his study of relationships": Yee, "The Psychology of Massively Multi-User Online Role-Playing Games," 196.

126 "According to a 2007 study": Dmitri Williams, Scott Caplan, and Li

Xiong, "Can You Hear Me Now? The Impact of Voice in an Online Gaming Community," *Human Communication Research* 33 (2007): 427–49.

126 "'Voice,' the authors maintain": Ibid., 444.

128 "'social trust is a valuable'": Putnam, *Bowling Alone*, 135.

129 "'wresting power from the few'": Lev Grossman, "Time's Person of the Year: You," *Time*, December 13, 2006.

129 "The entertainment business has begun": Bill Werde, "Philip Rosedale," *Billboard*, January 6, 2007, 26.

130 "In a 2008 interview": Dean Takahashi, "Q&A: Linden Lab CEO Mark Kingdon on Second Life's Latest Evolution," September 18, 2008, http://venturebeat.com/2008/09/18/qa-linden-lab-ceo-mark-kingdon-on-second-lifes-latest-evolution/.

SIX: Hooked

133 "'I am addicted to EQ'": Yee, "The Psychology of Massively Multi-User Online Role-Playing Games," 201.

134 "'People feel a lot of shame'": Pagan Kennedy, "Craft Addicts: Do Online Games Trigger a New Psychiatric Disorder?" *Boston Globe*, June 8, 2008, www.boston.com/bostonglobe/ideas/articles/2008/06/08/craft_addicts/.

134 "In a 2007 editorial": Jerald J. Block, MD, "Issues for DSM-V: Internet Addiction," *American Journal of Psychiatry* 165 (March 2008): 306–7.

135 "'About 86% of Internet addiction'": Ibid., 306.

136 "Throughout the remainder of the 1990s": For some of the early scholarly literature on Internet addiction, see Kimberly S. Young, "Psychology of Computer Use: XL. Addictive Use of the Internet: A Case That Breaks the Stereotype," *Psychological Reports* 79 (1996): 899–902; Viktor Brenner, "Psychology of Computer Use: XLVII. Parameters of Internet Use, Abuse and Addiction: The First 90 Days of the Internet Usage Survey," *Psychological Reports* 80 (1997): 879–82; Mark D. Griffiths, "Does Internet and Computer 'Addiction' Exist? Some Case Study Evidence," *CyberPsychology & Behavior* 3, no. 2 (2000): 211–18; and Chien Chou, Linda Condron, and John C. Belland, "A Review of the Research on Internet Addiction," *Educational Psychology Review* 17, no. 4 (December 2005): 363–88.

137 "'I.A.D.,' Goldberg explained": David Wallis, The Talk of the Town, "Just Click No," *New Yorker*, January 13, 1997, 28.

137 "'To medicalize every behavior'": Ibid.

137 "'Dependence on a chemical substance'": *The American Psychiatric Association's Psychiatric Glossary* (Washington, DC: American Psychiatric Press, 1984), 31.

138   " 'It also tells the memory' ": Eric J. Nestler Laboratory of Molecular Psychiatry, "Brain Reward Pathways," http://transmitter.neuro.mssm.edu/NestlerLab/paths_b02.php.

139   " 'these gender differences may help' ": Fumiko Hoeft et al., "Gender Differences in the Mesocorticolimbic System during Computer Game-Play," *Journal of Psychiatric Research* 42, no. 4 (March 2008): 253–58.

139   "We know from previous": M. E. Ballard and J. R. West, "Mortal Kombat™: The Effects of Violent Videogame Play On Males' Hostility and Cardiovascular Responding," *Journal of Applied Social Psychology* 26 (1996): 717–30; M. Flemming and D. Rickwood, "Effects of Violent Versus Nonviolent Video Games on Children's Arousal, Aggressive Mood, and Positive Mood," *Journal of Applied Social Psychology* 31 (2001): 2047–71.

140   "In a 2007 policy report": Mohamed K. Khan, "Report of the Council on Science and Public Health," CSAPH Report 12-A-07.

140   " 'video game overuse is most common' ": Ibid., 4.

140   " 'The APA does not consider' ": *Statement of the American Psychiatric Association on "Video Game Addiction"* (Arlington, VA: American Psychiatric Association, June 21, 2007).

140   " 'Revising DSM,' the APA concludes": Ibid.

140   " 'arguments, lying, poor achievement, social isolation' ": Block, "Issues for DSM-V: Internet Addiction," 306–7.

141   " 'a forum for partners, family, and friends' ": http://health.groups.yahoo.com/group/EverQuest-Widows/.

141   " 'We turn to each other' ": Ibid.

141   " 'Video games,' Orzack explains": Rob Wright, "Expert: 40 Percent of World of Warcraft Players Addicted," Tom's Games, August 8, 2006, http://us.tomsgames.com/us/2006/08/08/world_of_warcraft_players_addicted/.

141   "In one national survey": Yee, "The Psychology of Massively Multi-User Online Role-Playing Games," 196.

142   "In South Korea, a country where": For information on the rapid diffusion of broadband in South Korea, see *Subscribers of High-Speed Internet by Region* (Seoul, Korea: Ministry of Information and Communication, 2007); and Heejin Lee, Robert M. O'Keefe, and Kyoung-Lim Yun, "The Growth of Broadband and Electronic Commerce in South Korea: Contributing Factors," *Information Society* 19, no. 1 (2003): 81–93.

142   "In 1998 South Korea hosted": T. Park, "Analysis Report: Factors Leading to Sharp Increase Internet Users in Korea," Korean Network Information Center, 2000.

142 "The South Korean government estimates": Y. H. Choi, "Advancement of IT and Seriousness of Youth Internet Addiction," in *2007 International Symposium on the Counseling and Treatment of Youth Internet Addiction* (Seoul, Korea: National Youth Commission, 2007): 20.

142 " 'briefly stopped to smoke' ": Caroline Gluck, "South Korea's Gaming Addicts," BBC News World Edition, November 22, 2002, http://news.bbc.co .uk/2/hi/asia-pacific/2499957.stm.

142 "In 2006, ten South Koreans": Y. H. Choi, "Advancement of IT and Seriousness of Youth Internet Addiction."

143 " 'youngsters who become obsessed' ": Gluck, "South Korea's Gaming Addicts."

143 " 'It is most important' ": Martin Fackler, "In Korea, a Boot Camp Cure for Web Obsession," *New York Times*, November 18, 2007.

143 "Two researchers, Dal Yong Jin": Dal Yong and Florence Chee, "Age of New Media Empires: A Critical Interpretation of the Korean Online Game Industry," *Games and Culture* 3, no. 1 (January 2008): 38–58. See also Florence Chee, "Essays on Korean Online Game Communities" (master's thesis, Simon Fraser University, Canada, 2005).

144 " 'the PC Bang is the site' ": Florence Chee, "The Games We Play Online and Offline: Making Wang-tta in Korea," *Popular Communication* 4, no. 3 (2006): 225–39.

144 " 'A PC bang also has been known' ": Ibid.

144 "In one longitudinal study": Ducheneaut et al., "Building an MMO with Mass Appeal."

146 "In one national survey": Yee, "The Psychology of Massively Multi-User Online Role-Playing Games," 202.

146 "Fifty percent of those": Ibid.

147 "Research scientists have long": Robert J. Hancox, Barry J. Milne, and Richie Poulton, "Association of Television Viewing during Childhood with Poor Educational Achievement," *Archives of Pediatrics Adolescent Medicine* 159, no. 7 (2005): 614–18; Patricia A. Williams et al., "The Impact of Leisure-Time Television on School Learning: A Research Synthesis," *American Educational Research Journal* 19 (1982): 19–50. The literature on games and academic performance is more mixed. See Mary B. Harris and Randall Williams, "Video Games and School Performance," *Education* 105, no. 3 (1985): 306–9; and Vivek Anand, "A Study of Time Management: The Correlation Between Video Game Usage and Academic Performance Markers," *CyberPsychology & Behavior* 10, no. 4 (2007): 552–59.

147 "'Time spent with media'": "American Academy of Pediatrics: Media Education," *Pediatrics* 104, no. 2 (August 1999): 3, 341–43.

148 "We also believe that": Thomas Layton, "Games Cited in Recent Deaths," GameSpot, August 28, 2003, www.gamespot.com/news/6074253 .html.

149 "'For some people,' especially those": Castronova, *Synthetic Worlds*, 65.

149 "for others, perhaps the fantasy world": Ibid.

149 "making an understandable choice": Ibid.

149 "'He didn't want much'": Workbench: Programming and Publishing News and Comment, "Christina Cordell: 'A Survivor of Everquest Addiction,'" http://workbench.cadenhead.org/everquest/cordell1.html.

149 "'In a way,' she writes": Ibid.

150 "The players are acting": Ibid.

151 "I feel sorry that people": Ibid.

151 "Games researcher Nick Yee": Ibid., 198.

151 "For at least two decades": For a discussion of "tiny sex," see Turkle, *Life on the Screen*, 89.

151 "'is not only common'": Ibid.

151 "In his survey, Yee": Ibid.

151 "The 2006 Active Gamer Benchmark Study": Nielsen Entertainment, "Nielsen Entertainment Study Shows Video Gaming is Increasingly a Social Experience," news release, October 5, 2006, //www.prnewswire.com/cgi-bin/ stories.pl?ACCT=104&STORY=/www/story/10-05-2006/0004446115& EDATE.

152 "Though the data varies": For a discussion of women and games, see T. L. Taylor, *Play Between Worlds: Exploring Online Game Culture* (Cambridge, MA: MIT Press, 2006): 93–94.

152 "Overall it can be a cheaper": Ibid., 194.

153 "'First it [the computer] becomes'": David Smith, "Addiction to Internet 'Is an Illness,'" *The Observer*, March 23, 2008, www.guardian.co.uk/ technology/2008/mar/23/news.internet?gusrc=rss&feed=technology.

154 "'I still have a foot'": Workbench, http://workbench.cadenhead.org/ everquest/cordell1.html.

154 "'I'm going back to school'": Ibid.

SEVEN: Now!

157 "'For the younger generation'": Christine Rosen, "The Myth of Multi-tasking," *New Atlantis* no. 20 (Spring 2008): 105–10.

157 "'We now devour our pop culture'": Nancy Miller, "Minifesto for a New Age," *Wired*, March 2007.

158 "'When we first approached'": Steven Levy, *The Perfect Thing: How the iPod Shuffles Commerce, Culture, and Coolness* (New York: Simon & Schuster, 2006), 135.

160 "'Americans viewed more than'": *YouTube Draws 5 Billion U.S. Online Video Views in July 2008* (Reston, VA: comScore Media Metrix, September 10, 2008), www.comscore.com/press/release.asp?press=2444.

160 "Since the widespread diffusion": According to Nielsen Media Research, 29 percent of TV households in the United States owned a remote control. Less than ten years later, in 1994, more than 90 percent of TV households owned a remote control. Nielsen, *Television Audience: 1993*, 2.

162 "A 1972 study of southern California": J. Lyle and H. R. Hoffman, "Children's Use of Television and Other Media," in E. A. Rubinstein, G. A. Comstock, and J. P. Murray, eds., *Television and Social Behavior*, vol. 4, *Television in Day-to-Day Life: Patterns and Use* (Washington, DC: U.S. Government Printing Office., 1972): 129–256.

162 "In his study of media use by youth": Roberts and Foehr, *Kids and Media in America*, 42–48.

162 "Similarly, three-quarters of older children": Ibid., 42–43.

162 "'Compared with even a few years ago'": Ibid., 42.

163 "Children and teenagers spend at least": For a detailed assessment of media multitasking among young people, see Foehr, *Media Multitasking Among American Youth*.

163 "From their view of the world": Harris Interactive and Teen Research Unlimited, "Born To Be Wired: The Role of New Media for a Digital Generation."

165 "'The great irony of multitasking'": Walter Kirn, "The Autumn of the Multitaskers," *Atlantic*, November 2007, www.theatlantic.com/doc/200711/multitasking.

166 "To determine what happens": Paul E. Dux et al., "Isolation of a Central Bottleneck of Information Processing with Time-Resolved fMRI," *Neuron* 52 (December 21, 2006): 1109–20.

166 "'When humans attempt to perform'": Ibid., 1109.

167 " 'We are under the impression' ": Steve Lohr, "Slow Down, Brave Multitasker, and Don't Read This in Traffic," *New York Times*, March 25, 2007.

168 " 'When we talk about multitasking' ": Rosen, "Myth of Multitasking."

168 "we are motivated by a desire": http://continuouspartialattention.jot .com/WikiHome.

168 " 'is an always-on, anywhere, anytime' ": Ibid.

169 " 'in a 24/7, always-on world' ": Ibid.

169 " 'when people do their work' ": Rosen, "Myth of Multitasking."

EIGHT: "May I have your attention?"

171 "In a 2008 study of 2,089 mobile-phone users": Harris Interactive, "A Generation Unplugged: Research Report," September 12, 2008, 5.

172 "By 2007 a decisive majority": Nielsen Mobile, "Kids on the Go: Mobile Usage by U.S. Teens and Tweens" (New York: The Nielsen Company, December 6, 2007).

173 "First, the open policies acknowledge": Everett M. Rogers, *Diffusion of Innovations* (New York: Free Press, 1962).

173 "In a 2002 poll": John Horrigan, *Mobile Access to Data and Information* (Washington, DC: The Pew Internet & American Life Project, March 2008): 1.

173 "Mobile phones (57 percent) ranked ahead": Ibid., 1.

173 " 'are much more likely to say' ": Ibid., 6.

175 " 'They are used for everything' ": NBC News Forum, "Interview: New York City Schools Chancellor Joel Klein," September 2, 2007, www.nydaily news.com/blogs/dailypolitics/docs/misc/09020730.loc.doc.

175 " 'Kids are text messaging in their pockets' ": Ibid.

176 "A 2008 study found that": Harris Interactive, "A Generation Unplugged," 11.

176 " 'To you this is a tool' ": Matt Richtel, "Schools Relax Cellphone Bans, Nodding to Trend," *New York Times*, September 29, 2004.

176 "I feel naked": Julie Bosman, "Sweep at School Turns Up a Trove of Electronic Contraband," *New York Times*, June 1, 2007.

176 "In their e-mail messages to the school": Elissa Gootman, "Hardships of School Cellphone Ban Are Detailed in E-Mail Messages to Public Advocate," *New York Times*, November 15, 2006.

176 " 'The chancellor will have' ": Elissa Gootman, "City Schools Cut Parents' Lifeline (the Cellphone)," *New York Times*, April 27, 2006.

177 "Forty-two percent of the teens": Harris Interactive, "A Generation Unplugged," 12.

179 "In a 2007 study with Microsoft": Shamsi T. Iqbal and Eric Horvitz, "Disruption and Recovery of Computing Tasks: Field Study, Analysis, and Directions," in *Proceedings of the SIGCHI Conference on Human Factors in Computing Systems* (Workshop) (San Jose, CA: April 28–May 3, 2007).

179 "normal daily tasks that users perform": Ibid.

180 " 'Even when users respond immediately' ": Ibid.

184 "Can you repeat the question, please?": David Cole, "Laptops vs. Learning," *Washington Post*, April 7, 2007.

185 " 'The wall of vertical screens' ": Jeffrey R. Young, "The Fight for Classroom Attention: Professor vs. Laptop," *The Chronicle of Higher Education*, June 2, 2006.

186 " 'It's a condition induced by modern life' ": Alorie Gilbert, "Newsmaker: Why Can't You Pay Attention Anymore?" CNET News, March 28, 2005, http://news.cnet.com/Why-cant-you-pay-attention-anymore/2008–1022_3–5637632.html.

186 "is by definition the transmission": Don Tapscott, *Grown Up Digital: How the Net Generation Is Changing Your World* (New York: McGraw-Hill, 2008), 130.

186 " 'Sitting in front of a TV set' ": Ibid., 131.

186 " 'But unlike the entertainment world' ": Ibid.

188 "In a 1997 survey": Kathy Scherer, "College Life On-line: Healthy and Unhealthy Internet Use," *Journal of College Student Development* (November/December 1997).

189 "A Pew study reports": Deborah Fallows, *Browsing the Web for Fun* (Washington, DC: The Pew Internet & American Life Project, February 2006), 2.

189 " 'if you don't allow yourself' ": Gilbert, "Newsmaker."

CONCLUSION: A Message from Barack

193 " 'Young people are on the Web' ": Jose Antonio Vargas, "Young Voters Find Voice on Facebook," *Washington Post*, February 17, 2007.

193 " 'About how many people' ": www.youtube.com/watch?v=K_UZr3g-9yA.

194 " 'This is a remarkable' ": www.youtube.com/watch?v=i8tSonwApFE.

194 " 'About three weeks back' ": www.youtube.com/watch?v=K_UZr3g-9yA.

194 " 'I don't really understand it' ": Ibid.

194 "'Technology is changing'": Zachary A. Goldfarb, "Mobilized Online, Thousands Gather to Hear Obama," *Washington Post*, February 3, 2007.

194 "When Obama visited George Mason": Vargas, "Young Voters Find Voice."

195 "'discuss brands, companies, products, and services'": Tapscott, *Grown Up Digital*, 89. For a more complete discussion of what Tapscott refers to as "N-Fluence networks," see 192–201.

196 "'wanted Mr. Obama's social network'": Brian Stelter, "The Facebooker Who Friended Obama," *New York Times*, July 7, 2008.

196 "Sociologist Barry Wellman calls": Barry Wellman, "Physical Place and CyberPlace: The Rise of Personalized Networking," *International Journal of Urban and Regional Research* 25, no. 2 (2001): 227–52.

197 "'Just like Kennedy brought in'": Claire Cain Miller, "How Obama's Internet Campaign Changed Politics," *New York Times*, November 7, 2008.

197 "'were it not for the Internet'": Ibid.

198 "'to lower the cost of building'": David Carr, "How Obama Tapped into Social Networks' Power," *New York Times*, November 10, 2008.

199 "'It was like a guy'": Ibid.

199 "'Other politicians I have met'": Ibid.

199 "'was the first politician'": Ibid.

200 "'Thanks for sharing . . . all these'": www.flickr.com/photos/barack obamadotcom/3008255125/.

200 "'amazing openness'": Ibid.

200 "'Sorry—but I wouldn't'": Ibid.

200 "'Young voters,' Hart explains": Jann S. Wenner, "How Obama Won," *Rolling Stone*, November 27, 2008.

201 "Statistics from TubeMogul.com": Personal Democracy Forum: tech President.com, "YouTube Stats," http://techpresident.com/youtube/2008.

201 "TubeMogul estimated that the content": Micah L. Sifry, "How Much Is YouTube Worth to Obama and McCain?" October 24, 2008, www.tech president.com/blog/entry/32071/how_much_is_youtube_worth_to_obama_ and_mccain.

201 "'the finer point would be'": Ibid.

201 "According to the *Nation*'s Ari Melber": Ari Melber, "YouTubing the Election," the *Nation*, November 4, 2008.

202 "'a sort of metropolitan idea'": www.onthemedia.org/transcripts/ 2007/11/02/07.

202 "'the expansive commentary is fast'": Virginia Heffernan, "God and Man on YouTube," *New York Times Magazine*, November 4, 2007.

203 "'participation in presidential elections'": Robert Putnam, *Bowling Alone*, 32.

203 "'In 1973 most Americans'": Ibid., 44.

203 "'*most did not engage*'": Ibid., 44.

203 "'Those activities that brought'": Ibid., 45.

204 "Pew's findings may surprise": Scott Keeter, Juliana Horowitz, and Alec Tyson, *Young Voters in the 2008 Election* (Washington, DC: Pew Research Center for the People & the Press, November 12, 2008), http://pewresearch.org/pubs/1031/young-voters-in-the-2008-election.

205 "In 1987 the Roper Organization": The Roper Organization, *America's Watching*.

205 "In a December 2008": *Internet Overtakes Newspapers as News Outlet* (Washington, DC: Pew Research Center for the People & the Press, December 23, 2008), http://pewresearch.org/pubs/1066/internet-overtakes-newspapers-as-news-source.

205 "'Nearly six-in-ten Americans'": Ibid.

206 "Tufts University's Center for Information": The Center for Information & Research on Civic Learning and Engagement, "Youth Voting," www.civicyouth.org/?page_id=241.

206 "One thing that's been incredibly clear": Goldfarb, "Mobilized Online, Thousands Gather to Hear Obama."

207 "'elections have not been'": Andy Oram, "Don't Say the Internet Has Changed Elections," November 4, 2008, http://broadcast.oreilly.com/2008/11/dont-say-the-internet-has-chan.html.

207 "'If different generations have different tastes'": Putnam, *Bowling Alone*, 34.

absence-in-presence, 48
ADD (attention deficit disorder), 186
addiction. *See* Internet addiction
ADT (attention deficit trait), 186
adults, 7–8, 22, 38–40, 91. *See also*
  parents
African Americans, 8, 31–33, 42–43,
  76, 78
AIM (AOL Instant Messenger), 4,
  5, 163
ambient intimacy, 56
American Academy of Pediatrics, 50,
  147
American Library Association (ALA),
  21
American Medical Association
  (AMA), 139–40
American Psychiatric Association
  (APA), 134–35, 137–40
Anderson, Chris, 45
Anderson, Tom, xii
Andreessen, Marc, 198–99

AOL Instant Messenger (AIM), 4,
  5, 163
Asian Americans, 78, 116
attention. *See* continuous partial
  attention (CPA)
attention deficit disorder (ADD), 186
attention deficit trait (ADT), 186
avatars, 11, 100–101, 113, 114,
  115–21

Bailenson, Jeremy, 119–20
Battelle, John, xiii
Big Sort, 97–98
Bishop, Bill, 97–98
blacks. *See* African Americans
Block, Jerald J., 134–35, 140–41,
  153–54
blogging, xiii, 11, 76, 90, 115, 118,
  180
Bloomberg, Michael, 175
bonding social capital, 95–97. *See also*
  community; friendships

Bourdieu, Pierre, 82–83
boyd, danah, 75–76
brain research, 138–39, 148, 165–67
bridging social capital, 95–96
broadband, 8–11, 33, 88, 210
Brown, John Seely, 121
Brown, Merrill, x
Brynjolfsson, Erik, xx

Caplan, Scott, 126
Castronova, Edward, 112–13, 149
cell phones. *See* mobile phones
Chee, Florence, 143–44
children, 6, 7, 14–15, 32, 50, 78–79, 162
citizenship. *See* digital citizenship; politics
class divisions, xx, 29–35, 37, 75–100
Clinton, Bill, 30–31
code-switching, 94–95
community, xviii–xix, 51, 53–54, 63–64, 122–23, 127–28, 196–97. *See also* friendships
computers. *See* digital media; Internet; laptops; libraries; schools; *and specific digital media*
comScore Media Metrix, 38, 70, 104–5
continuous partial attention (CPA), 168–69, 172, 183–86
convergence culture, 115
Cordell, Brianna and Christina, 148–55
CPA (continuous partial attention), 168–69, 172, 183–86
cultural capital, 82–83, 88
cyberbalkanization, xx

cyberbullying, 33–35
cyberpredators, 20–24, 36–37, 65

Dean, Howard, 195–96
Delete Online Predators Act (DOPA), 20–24
DeWolfe, Chris, xii
diffusion of innovations, 173
digital cameras, 41, 45–46, 175
digital citizenship, 29
digital divide, 30–35, 76. *See also* class divisions; race
digital media: children's use of, 6, 7, 14–15, 50; compared with television, 8–18, 61, 99–100, 115, 186, 205; hours spent with, 50, 61–62, 104, 161–62; limits and difficulties produced by, xx, 47–52, 144–51, 187–89; research methods on, xiv, 209–16; and wired castle, 2, 161. *See also* class divisions; gender; Internet; race; schools; *and specific media*
displacement theory, 63
do-it-yourself (DIY) media, 10–11, 82
DOPA (Delete Online Predators Act), 20–24
Ducheneaut, Nicolas, 114

Eastwick, Paul, 100–101
education. *See* libraries; schools
Education Department, U.S., 31, 32
Ellison, Nicole, 68, 94
e-mail, 4, 6, 7, 25, 40, 84, 179, 180, 184, 199–200
entertainment. *See* fast entertainment; television

e-stalking, 56
Evans, Beth, 24–27
EverQuest, 141, 144, 148–51, 154

Facebook: and addiction debate,
    133–34; adults' use of, 8, 38–40,
    91; aesthetics of, 80–83; creation
    of, 196, 210; demographics of,
    38–40, 80, 83–86, 91, 96–97; as
    distraction from academic work,
    183, 184, 187–89; and events in-
    vitations, 62–63; and friendships,
    66–68, 71–74, 122; and groups,
    26; legislation on, 20; motivations
    for and uses of, xv, xvi, 5, 46, 70;
    MySpace versus, for college stu-
    dents, xx, 70, 77–91, 94, 99–100;
    news-feed feature of, 190; and
    Obama campaign, 193–95, 197,
    203, 204; opening up registration
    in, 38, 84, 87, 90–91; parents'
    monitoring of, 36–37; and privacy,
    85–87; profile in, 118–19; and
    race and class divisions, 75–100;
    safety of, and security issues on,
    36–37, 85, 87–88; schools' block-
    ing of, 27–28, 93; statistics on, xi,
    xv, 38, 91, 190; teachers' use of,
    38–40; use of, and time spent on,
    ix, xiv, xv, xx, 17, 35–36, 187, 189,
    190; and video file-sharing, 158
"Face-in-the-Door" condition,
    100–101
families. *See* intimacy; parents
fast entertainment, 157–61, 172,
    190–91
Fey, Tina, 204
FIM (Fox Interactive Media), x–xi

Fischer, Claude, 49, 51–52, 54–55
Fitzpatrick, Mike, 20–23
Flickr, 11, 197, 200
Fox Interactive Media (FIM), x–xi
friending, xvi–xvii, 39–40, 54, 88
friendships: of gamers, xvii–xviii,
    61–62, 66–68, 70–73, 107–14,
    122–29; online versus face-to-
    face friendships, 53–54, 60–62,
    127, 131; and phatic interactions,
    71–73; and social isolation due to
    digital media, 47–52; and social-
    network sites, 62–74, 96–97, 122.
    See also intimacy
Friendster, xii, xvi, 5, 20

Gardner, Wendi, 100–101
gated communities, 86–87
Gee, James Paul, 183
gender: and digital media, 6–7, 10,
    78, 104–6; of gamers, 105–6,
    124, 139; and gender swapping,
    116–18; and telephone, 49; and
    television, 103–4
gender swapping, 116–18
Gibbs, Martin R., 71–72
Goffman, Erving, 46
Goldberg, Ivan, 135–37
Granovetter, Mark, 68–69
guilds, 121, 122, 124–29, 145, 153
Guitar Hero, 108–10, 197

Hallowell, Edward, 186, 189
Halo, 108–10, 126
Hansen, Chris, 23
Hargittai, Eszter, 78–79
Harris Interactive, 172, 177
Harrison, Phil, 110

Hart, Peter, 200
health issues, 50, 142–44, 145, 146.
    See also Internet addiction
Heffernan, Virginia, 202
Hispanics. *See* Latinos
HomeNet Project, 6–8
Horvitz, Eric, 179–80
Howard, Steve, 71–72
Huffington, Arianna, 197–98
Hughes, Chris, 196
Hurley, Chad, 159–60

IAD (Internet addiction disorder),
    134–37, 139
IGN Entertainment, Inc., x–xi
IM. *See* instant messaging (IM)
income. *See* class division
information disadvantaged, 30
instant messaging (IM), xiii, 4–6,
    8, 25, 47–49, 56–57, 59, 72–73,
    179–80
intercohort, 207
Internet: age of users of, 5, 7–8,
    14–15, 104; attitudes on, 16–17,
    187; broadband for, 8–11, 33,
    88, 93, 210; as distraction from
    homework, 187–89; gender of
    users of, 6–7, 104–6; hours spent
    on, 104–5; as human network,
    55–56; and multitasking, 163–65;
    as news source, 205; and social
    capital, 63–64; social connection
    as "holy grail" of, 52–53; statis-
    tics on, x, xxi, 3, 5, 8, 10, 12, 20,
    31, 61, 104–5, 173, 189, 205. *See
    also* class divisions; digital media;
    e-mail; libraries; race; schools;
    and specific digital media

Internet addiction, 133–55
Internet addiction disorder (IAD),
    134–37, 139
intimacy, xix, 52, 56, 74, 99–100,
    119–20, 147–48, 151–55. *See also*
    sexuality
iPhones, 158–59, 161, 171
iPods, 47, 157–59, 161, 172, 174,
    175, 177–80
Iqbal, Shamsi T., 179–80
iTunes, 158, 161, 172

Jenkins, Henry, 115
Jin, Dal Yong, 143–44
Jobs, Steve, 158
Justice Department, U.S., 22

Kaiser Family Foundation, 15, 50,
    163
Kirn, Walter, 165
Kraut, Robert, 6–7

Lampe, Cliff, 68
laptops, 183–86
Latinos, 8, 31–33, 42–43, 76–78,
    92–94
Levinsohn, Ross, xi
Levy, Steven, 158
libraries, 24–27, 32, 93, 180–81
life sharing, xvii, 64, 70, 131
LinkedIn, 8, 70, 118, 197
Low, Setha, 86–87

MacArthur Foundation, xiv, 24
Marois, René, 165–67
McCain, John, 201
Microsoft, xvii–xviii, 110, 179
minisodes, 159, 161

MMOGs (massively multiplayer online games), 130–31

MMORPGs (massively multiplayer online role-playing games), 106, 108, 113, 115–21, 123–24, 133, 134, 141, 144–46, 151–53

mobile phones, 28, 33, 44, 47–48, 56, 57, 72, 157, 171–79

MUDs (multiuser dungeons), 7, 116–17

multitasking, 161–69, 183–86

multiuser dungeons (MUDs), 7, 116–17

Murdoch, Rupert M., ix–xi, xiii, 19, 91–92

music and music videos, 10, 43, 105, 158–64, 172, 201–2

MySpace: and addiction debate, 133–34; adults' use of, 8, 22, 38–40; aesthetics of, 80–83, 99; creation and rise of, xi–xiii, 210; demise of, 92; demographics of, xi, 8, 22, 38–40, 80, 83–88, 90, 96–97; e-mail versus, 25; and "emo" teens, 90, 91; Facebook versus, for college students, xx, 70, 77–91, 94, 99–100; features of, xiii–xiv; and friendships, 66; and identity work, 42–43; legislation on, 20; and libraries, 24–27; media coverage of, 58; motivations for and uses of, xv–xvi, 5, 46, 58–59, 70; News Corporation's acquisition of, xi, xiii, 19, 91–92; and Obama campaign, 197; open-door policy of, 87–88, 91, 92; parents' monitoring of, 36–37; profile in, 11, 46; and race and class

divisions, 75–100; reasons for avoiding, xv–xvi; schools' blocking of, 27–28, 93; security issues on, 36–37; and sexual and gender identities, 42–45; statistics on, xi, 38, 91; teachers' use of, 38–40; use of, and time spent on, xv–xvi, 17, 35–36, 58, 188; and video file-sharing, 158

National Telecommunications and Information Administration (NTIA), 30, 32

NBC Universal, 13, 209–10

Negroponte, Nicholas, x

Nestler, Eric, 138–39

Netville study, 63–64

Nielsen Entertainment, 105–7, 151–52

Nielsen Media Research, xi, 9, 91, 103–4, 160, 210

NTIA (National Telecommunications and Information Administration), 30, 32

Obama, Barack, 98–99, 193–208

Oldenburg, Ray, xix, 58–59, 129

Orzack, Maressa Hecht, 141

parents, 6, 7, 32, 35–37, 47–52, 78–79, 174, 176, 178. See also adults

participation divide, 32

PC Bangs, 142–44, 154

personal computers. See Internet; laptops

Perverted Justice, 23

Pew Internet & American Life Project, xi, 5, 8, 20, 23, 32–34, 56–57, 65, 85, 93, 105, 173, 210

Pew Research Center, 10, 11, 16, 36, 189, 204, 205

phatic interactions, 71–73

phones. *See* iPhones; mobile phones; telephones

photographs, 41, 44, 45–46, 175, 200

podcasts. *See* iPods

politics, 98–99, 193–208

Poltrack, David F., 103–4

predators. *See* cyberpredators

Prescott, LeeAnn, 14

presence-in-absence, 48

privacy, 45–46, 85–87

Proteus Effect, 119–21

Putnam, Robert, xviii–xix, 8, 53, 54, 95, 203, 207

race, xx, 8, 31–33, 42–43, 75–101

radio, online, 10

Rheingold, Howard, 45, 49, 53

Rice, Stuart, 52

Rideout, Victoria, 15

Roberts, Donald F., 162

Rogers, Everett M., 173

role-playing games. *See* games; MMORPGs (massively multiplayer online role-playing games)

Roper Organization, 15–16, 205

Rosen, Christine, 157, 168

Rospars, Joe, 196

Rushkoff, Douglas, 104

safety. *See* security and safety issues

Schement, Jorge, 2, 161

Scherer, Kathy, 188

Schiesel, Seth, 111

schools: ban on digital media in, 27–28, 93, 174–78, 184–85; computers in elementary schools, 14; and continuous partial attention (CPA), 183–86; and cyberbullying, 33–35; digital media used by educators in, 180–83; discipline by, 40–42; distractions from digital media in, 145–46, 183–86, 187–89; lectures in, 186; and mobile phones, 172–79; race and class divisions in, 29–37, 31; SMART Boards in, 181–82; and social-network sites, 28–46; teachers' attitudes toward personal media in, 177–79; teachers' use of social-network sites, 38–40; and teaching about digital world, 28–29, 35–37

Second Life, 120, 129–30

security and safety issues, 35–37, 85–88

sexting, 44

sexuality, xviii, 43–45, 50, 151

sexual predators. *See* cyberpredators

short message service (SMS), 57

SMART Boards, 14, 181–82

SMS (short message service), 57

social capital, xix, 35, 37, 53, 64, 74, 95–97, 129

social class. *See* class divisions

social networks, 68–69. *See also* community; friendships

social trust, 127–28

Sony, 110–11, 159

South Korea, 142–44, 154

space-transcending technologies, 51–52

stalking. *See* e-stalking

Steinfield, Charles, 68
Stelter, Brian, 196
Stone, Linda, 168–69

Tapscott, Don, 13, 186, 195
telephones, 49, 52, 54–55, 173. *See also* iPhones; mobile phones
television: attitudes toward, 15–17; cable television, 2; for children, 14, 162; compared with digital media, 8–18, 61, 99–100, 106–7, 115, 186, 205; and consumer culture, 161; on cyberpredators, 22–23; impact of, on social relationships, xviii–xix; and Kennedy presidency, 197; male viewership of, 103–4; and multitasking, 162–64; as news source, 205; number of channels received in U.S. homes, 160; and parents' educational attainment, 78; and race, 99–100; sex and violence on, xviii, 44; statistics on, 1–3, 8–9, 13, 103, 160, 162, 173, 189, 205, 210; time spent viewing, 8, 9, 61
text messaging, 47–48, 57–58, 72–74, 167, 175, 177, 199–200
There.com, 100–101
third places, 58–59, 64, 127, 131
Thomas, Douglas, 121
Thompson, Clive, 56, 67
Time, ix, xiii, 1, 19, 129, 210
Trippi, Joe, 197
trust, 127–28
Turkle, Sherry, 116–17, 151
TV. *See* television
Twitter, 56, 67, 197, 199, 203

Van Alstyne, Marshall, xx
Vandewater, Elizabeth, 15
Vetere, Frank, 71–72
victimization of youth, 20–24
video file-sharing, 158–60, 182, 201–2, 204
violence in media, xvii, 50
virtual cross-dressing, 116–18
Voice over Internet Protocol (VoIP), 124–26
voter turnout, 206–7. *See also* politics

Wartella, Ellen, 15
Web. *See* digital media; Internet
Wellman, Barry, 63–64, 70, 196
Wii, xviii, xix, 61, 110–12
wikis, 76, 115
Wiley, Malcolm, 52
Williams, Dmitri, 126
World of Warcraft (WoW), 108, 112–15, 117–19, 121–23, 125–29, 133, 139, 141, 144–45, 147, 152–54
Wurtzel, Alan, 103, 104

Xbox 360, xvii–xviii, 110
Xiong, Li, 126

Yahoo!, 158, 159
Yee, Nicholas, 119–21, 146, 151
YouTube, 158–60, 182, 183, 197, 199–204, 210

Zimmerman, Eric, 107
Zogby, John, 98
Zucker, Jeff, 13